MODEL
BOILER-MAKING

A Practical Handbook on the Designing, Making, and Testing of Small Steam Boilers

BY

E. L. PEARCE

REVISED BY

V. W. DELVES-BROUGHTON

FULLY ILLUSTRATED

First published 1920
Reprinted 1977
T.E.E. Publishing © ISBN 0 905100 13 1

£1.50

PREFACE

OWING to the number of queries on Model Boiler Construction received from readers of *The Model Engineer*, it was suggested to me by the Editor that a small book for amateur model makers explaining the rules for calculating the proportions and giving the methods of construction of model boilers would be useful, as the literature on this subject is very scanty. In carrying out this suggestion, I have endeavoured to give several examples of the different types of boiler in use, together with alternative methods of construction, showing how flanging may be dispensed with to a certain extent, and indicating where it is necessary to depart from the usual design of full-sized boilers. In a marine boiler, for instance, having return tubes, the combustion chamber from which the tubes start would be inside the boiler, with a water space at the back and sides; but in a model, owing to the restricted space, it would be a very difficult matter to follow this arrangement. Similarly, boilers of the locomotive type having over 200 small tubes would, if reproduced to a small scale, have tubes so small in diameter as to be useless, so that a few tubes of comparatively large diameter are substituted. The usual fault in amateur-made boilers is either that the boiler is not large enough for its work, or, if large enough, sufficient effective heating surface is not provided. Both these defects will be avoided if the principles explained in the following pages are borne in mind.

The designs I have given are all to scale, and may therefore be used as actual working drawings if they happen to suit the reader's particular requirements in point of size. If not, a consideration of the chapter on calculating proportions will enable the reader to adapt any of the designs to suit a larger or smaller engine than the one for which it was originally intended.

E. L. PEARCE.

Made and Printed in Great Britain.

PREFACE TO THE NINTH EDITION

SINCE this book was last revised, the progress in boiler making, like all other branches of model engineering, has advanced to such an extent as to necessitate an entire rearrangement of the matter contained therein.

Whilst descriptions and designs of typical boilers have been retained and brought up to date, much that was obsolete has been removed to make room for instructions which, being given in the simplest possible language, will, it is hoped, enable the budding engineer to design for himself, with safety, a boiler which will exactly fill his momentary requirements.

It is a much more worthy performance to design an appliance or apparatus for a definite purpose than to slavishly copy a design, which may be approximately what is required, simply because it happens to appear in a book.

V. W. DELVES-BROUGHTON.

MODEL BOILER MAKING

CHAPTER I

GENERAL PRINCIPLES OF BOILER DESIGN, MATERIALS, SHAPE, PROPORTIONS, STRENGTH, CAPACITY, HEATING SURFACE

In designing model and small steam boilers the main points to be observed are, to give the various parts the best shape and thickness to resist the pressure of steam the boiler is intended to carry ; to provide sufficient heating surface to raise the quantity of steam required ; to allow the proper amount of water and steam space, and to make suitable firing arrangements.

As most model boilers are worked at comparatively low pressures, the water level must not be carried too high, otherwise " priming " (*i.e.* the carrying over of water with the steam into the engine cylinder) will take place. Steam at a pressure of 25 lbs. per sq. in. requires a larger surface to rise from than an equal weight of steam at 100 lbs. per sq. in., and this is why, in many cases, boilers fail to steam satisfactorily when they are " over-cylindered." The large cylinders lower the pressure considerably ; and owing to the small surface from which the steam rises the reduction of pressure is attendant with violent ebullition, which causes priming and the comparative failure of the boiler as a steam generator for the given engine.

Model boilers are also often required to work for as long a period as possible without replenishing the water, not being fitted with a continuously working feed pump, so that it is better to give them a slightly larger water capacity than is generally allowed in larger boilers, or, which comes to much the same thing, to apparently " over boiler " the engine.

The materials that may be used are copper, stout charcoal

GENERAL PRINCIPLES OF BOILER DESIGN, ETC.

plate, and mild steel. Copper has the advantage of being more easily worked, brazed, or soldered, and is not subject to corrosion. Generally speaking, however, the use of tin plate or anything of the nature of iron or steel under $\frac{1}{8}$ or $\frac{5}{32}$ in. in thickness is to be deprecated.

The shape of a boiler is determined by the position it is to occupy and the work it is required to do. Wherever possible it should be made circular, that being the best form for resisting pressure. Any deviation from the circular form should be stayed or tied to an opposite point, such as the flat end plates in Cornish or Lancashire boilers, the crown plates of shell and firebox in vertical boilers, and the rectangular fireboxes of locomotive boilers, and boilers of the marine type having flat sides and top to fit the hulls of steamers. These flat-sided marine boilers are now practically obsolete, owing to the increased steam pressure now obtaining both in large and model practice.

The strength of a boiler is the most important consideration. For a given thickness and diameter a piece of solid drawn tube will make the strongest boiler. When the barrel or other cylindrical shell is made from a sheet of metal with the joint lapped and riveted, the strength depends upon the kind of joint made. *Theoretically*, a double-riveted butt joint with cover straps on each side of the plate will give a joint equal in strength to 75 per cent. of the original plate (see double butt joint, fig. 10).

Double-riveted lap joints (fig. 7) equal 66 per cent. Single-riveted lap joints only have a theoretical strength of 55 per cent.

Besides this, however, workmanship has to be taken into account, as the above figures only hold good when the holes are drilled perfectly fair, have the sharp corners chamfered off, and the rivets driven perfectly by hydraulic riveter.

For this reason both the Board of Trade and Lloyd's Agency insist upon a factor of safety of 5, with certain additions relative to the imperfections in workmanship. Thus if a boiler has its rivet holes punched before bending the plates, ·75 has to be added to the factor of safety.

Most reliable makers do not work to such a close margin, however, and, instead of calculating these items, put the best possible work into a boiler and still allow a factor of safety of 8, 10, or even 12. To make this quite clear, it should be here stated that a factor of safety means that after calculating the pressure necessary to burst the boiler, the working pressure is found by dividing this figure by the factor of safety, which should never be lower than 8, and there is no harm in allowing as much as 12, as even quite small boilers are capable of an enormous amount of destruction if they *do* happen to explode.

The following tables of riveting give the best *practical* dimensions for rivets, pitch, etc., for different thicknesses of plate. It makes no difference what the plate consists of, provided that the rivets are of the same material.

Single-riveted Lap Joints.

Thickness of plate.	Diameter of rivet.	Lap.	Pitch of rivets.	Percentage of strength of plate.
$\frac{3}{16}$ in.	$\frac{7}{16}$ in.	$1\frac{3}{16}$ in.	$1\frac{1}{2}$ in.	55·0
$\frac{5}{32}$,,	$\frac{3}{8}$,,	$1\frac{3}{32}$,,	$1\frac{3}{32}$,,	54·8
$\frac{1}{8}$,,	$\frac{9}{32}$,,	$\frac{3}{4}$,,	$\frac{23}{32}$,,	54·4
$\frac{3}{32}$,,	$\frac{7}{32}$,,	$\frac{9}{32}$,,	$1\frac{1}{16}$,,	55·0
$\frac{1}{16}$,,	$\frac{9}{64}$,,	$\frac{3}{8}$,,	$\frac{3}{8}$,,	54·6

Double-riveted Lap Joints.

Thickness of plate.	Diameter of rivet.	Pitch of rivets.	Lap of plates.	Distance between rows of rivets.	Percentage of strength of plate.
$\frac{3}{16}$ in.	$\frac{3}{8}$ in.	$1\frac{1}{2}$ in.	$1\frac{7}{8}$ in.	$\frac{5}{8}$ in.	66
$\frac{5}{32}$,,	$\frac{5}{16}$,,	$1\frac{1}{4}$,,	$1\frac{1}{2}$,,	$\frac{9}{16}$,,	66
$\frac{1}{8}$,,	$\frac{1}{4}$,,	$\frac{15}{16}$,,	$1\frac{3}{16}$,,	$\frac{7}{16}$,,	66
$\frac{3}{32}$,,	$\frac{3}{16}$,,	$\frac{3}{4}$,,	$\frac{15}{16}$,,	$\frac{5}{16}$,,	66
$\frac{1}{16}$,,	$\frac{1}{8}$,,	$\frac{15}{32}$,,	$\frac{9}{16}$,,	$\frac{7}{32}$,,	66

Double-riveted Butt Joint with Double Cover Straps.

Thickness of plate.	Diameter of rivet.	Pitch of rivets.	Width of straps.	Distance between rows of rivets.	Thickness of straps.	Percentage of strength of plate.
$\frac{5}{16}$ in.	$\frac{13}{32}$ in.	2 in.	$4\frac{1}{8}$ in.	$1\frac{3}{16}$ in.	$\frac{1}{4}$ in.	75
$\frac{1}{4}$,,	$\frac{3}{8}$,,	$1\frac{3}{4}$,,	$3\frac{5}{8}$,,	$1\frac{1}{16}$,,	$\frac{3}{16}$,,	75
$\frac{3}{16}$,,	$\frac{7}{32}$,,	$1\frac{1}{16}$,,	$2\frac{1}{4}$,,	$\frac{7}{16}$,,	$\frac{9}{64}$,,	75
$\frac{5}{32}$,,	$\frac{13}{64}$,,	1 ,,	$2\frac{1}{8}$,,	$\frac{13}{32}$,,	$\frac{1}{8}$,,	75
$\frac{1}{8}$,,	$\frac{3}{16}$,,	$\frac{7}{8}$,,	$1\frac{15}{16}$,,	$\frac{11}{32}$,,	$\frac{3}{32}$,,	75

GENERAL PRINCIPLES OF BOILER DESIGN, ETC.

The flat ends of a boiler should be made about 25 per cent. thicker than the shells; thus if the shell be $\frac{3}{16}$ in. thick, the ends should be $\frac{1}{4}$ in. thick. Tube plates must be still thicker, especially in model boilers, where the size of the tubes is large in proportion to the size of the boiler, a good rule being $1\frac{1}{2}$ the thickness of the shell plate, or $\frac{1}{4}$ the diameter of the tube, *whichever happens to be the thicker*.

Tube plates must be stiffened by stays like all other flat surfaces, but the tubes themselves can be made to serve this purpose if they are expanded into the tube plates and the ends carefully beaded over to prevent all tendency to draw out.

Another way of fixing stay tubes is to cut a fine thread on each end, and fix them into the tube plates with double lock nuts as shown in fig. 11.

Brazing or silver-soldering should not be depended upon for staying purposes where there is any chance of the heat acting directly on the brazed joint, without the latter being surrounded with water as the upper ends of the tubes in figs. 10 and 11.

Any flat surface exceeding 20 times the thickness of the plate in diameter or width should be stayed by stays pitched at 16 times the thickness of the plate apart. Thus supposing the boiler to be 4 in. in diameter, and the end plate $\frac{1}{8}$ in. thick, a single stay in the middle would be sufficient, and this stay would have to be strong enough to bear the stress due to the pressure of steam acting on an imaginary disc., 2 in. in diameter, with a factor of safety similar to that determined upon for the rest of the boiler.

If the boiler were 8 in. in diameter, however, and the ends still $\frac{1}{8}$ in. thick, seven stays would be required, one in the centre, and six spaced equally round a circle drawn round the centre with a radius of 2 in. Each stay will still have to support a stress due to the pressure of steam acting on a circle of 2 in. in diameter, as already explained.

There are only three metals which should be used in boiler construction, and these are mild steel, charcoal or Lowmore iron, and copper. Tinned iron should never be used, as it is very liable to corrosion; and brass should be avoided, as it is liable to spontaneous deterioration, besides which " drawing cracks " are often formed in the process of rolling, and are difficult to detect, as they often do not develop till the boiler has been heated and cooled several times.*

Mild steel is, however, a very vague term, and may have a tensile strength ranging anywhere between 20 and 35 tons per sq. in., and varying in nature from nearly pure iron to steel capable of being hardened. As the model engineer cannot have each plate that he uses tested, he must trust the merchant from whom he purchases his materials to supply him with suitable boiler plate, and not trust to fancy brands such as various crowns and XX's with B's underneath, as BBB as often stands for Beastly, Bad, Belgian, as for Best, Best, British.

A good test that anyone can try for himself is to take two strips of the metal about an inch wide, cut off two adjacent sides of the plate, and bend them, cold, over on themselves, when the strip should be capable of being turned right back upon itself (hairpin fashion), making a radius at the bend of not more than half the thickness of the plate, without showing any signs of cracking. This test will ensure the plate being tough, both in the direction in which it has been rolled and transversely—a more important attribute than great tensile strength.

For all practical purposes, steel suitable for boiler plates may be taken as having an ultimate tensile strength of 23 tons or 50,000 lbs. per sq. in.

Sheet copper is now so universally good that it is difficult to obtain a bad sample, and the only precaution necessary in purchasing is to closely inspect each sheet for surface flaws and blisters, which can often be detected by feel where not observed by a careful examination with a magnifying glass.

The tensile strength of copper may be taken as 25,000 lbs. per sq. in. ; the strength of sheet copper will be found in textbooks to be given as 27,500 lbs. per sq. in., but in boiler work this amount cannot be allowed, as the strength of copper rapidly diminishes as the heat to which it is subjected is increased, and for very high pressures, where the steam is consequently hotter, even less than 25,000 lbs. should be counted

---

*Note.—A brass disc which had stood being flanged and drilled, and being put to considerable rough usage in the process, on being silver-soldered into the end of a container, flew into two pieces, although it was, as far as could be judged by close inspection, thoroughly sound.

GENERAL PRINCIPLES OF BOILER DESIGN, ETC. 11

upon, especially in flue plates or tubes on which the powerful flame of a modern blow-lamp plays directly.

Charcoal iron is an excellent material, easy to work and less liable to corrosion than steel, but it is difficult to obtain now sufficiently thick for boiler making.

Lowmore iron can still be obtained, however, and this is as good for all practical purposes, and has the advantage that it is easier to flange and forge, especially for unskilled boilersmiths, than steel. The tensile strength of this can be taken at 20 tons or 45,000 lbs. per sq. in.

Now, having discussed the materials of which the boiler can be constructed, and the relative strength of the riveted joints of three types, we must next go on to the consideration of the strength of boilers as a whole. For this purpose, let us take a plain cylindrical boiler of 4 in. in diameter, made of $\frac{1}{16}$-in. sheet copper 8 in. long with $\frac{5}{64}$-in. flanged copper ends, the longitudinal seam to be lapped and double riveted.

First take the barrel or shell into consideration.

The steam acting upon the inner surface will tend to tear the plate along two imaginary lines running along the shell from end to end diametrically opposite one another—thus dividing the stress into two equal parts. The diameter of the boiler multiplied by its length will give us the area subject to steam pressure tending to cause this rupture, in this case 32 sq. in.; against this we have two strips of metal, $\frac{1}{16}$ in. thick by 8 in. long, helping to resist this rupture.

So we can take half the area of 16 sq. in. subject to steam pressure tending to burst $\frac{1}{16} \times 8$ in. $= \frac{1}{2}$ sq. in. of copper. We have already stated that copper requires 25,000 lbs. per sq. in. to cause its breakage, so the pressure on the boiler to burst the plate will have to be $\frac{25000}{2} = 12,500$ lbs., which divided by 16 (the number of square inches subject to steam pressure acting on each side of the boiler)=or 780 lbs. per sq. in.

On referring to the table of double-riveted joints it will be seen that the strength of the seam is ·66 that of the plate, $780 \times ·66 = 514$ lbs., the strength of the weaker point, *i.e.* the longitudinal seam.

The flanged ends will only require single riveting, as the pressure tending to force the end off is equal to the area of the end in square inches multiplied by the pressure to which it is subjected.

The area of the end is equal to πr^2, or $3·14 \times 2 \times 2 = 12·56$

sq. in., which multiplied by 514 (the pressure necessary to rupture the longitudinal seam), is equal to 6455 lbs. distributed along the end of the shell plate, which will measure $4 \times 3.14 = 12.56$ in. long by $\frac{1}{16}$ in. thick* $= .78$ of a sq. in., which at 25,000 lbs. per sq. in. (the tensile strength of copper) $= 19,500$ lbs., which in turn, multiplied by the percentage allowed for a single-riveted joint (see table), $.54 = 10,530$ lbs.—very nearly double the pressure to which it can be subjected.

It will thus be seen that if the longitudinal seam is double riveted, the ends and all circumferential seams need only be single riveted, and this rule holds good from the very smallest to the largest of boilers.

The end will also require to be supported by stays, for although the circumferential seam is amply strong enough to bear the strain, the pressure acting on the flat plate would cause it to buckle out of shape and put undue cross strains on the seam.

Through stays from end to end of the boiler are the simplest solution of the difficulty, but there are several other methods capable of being employed, at any rate in large boilers where a man can enter to "hold up" so that the riveting can be properly executed, but in models it is advisable not to employ such complications, as the weight saved is small, and the difficulty of getting at such awkward corners as occur in riveting " gusset " plates, etc., militates against good work.

As already explained, seven through stays should be employed, each capable of sustaining the stress induced by a pressure of 514 lbs. acting on the area of a circle 1 in. in diameter, *i.e.* $514 \times .78 = 390$ lbs.

It is advisable to use copper rods for the stays in a copper boiler, as it is not wise to employ dissimilar metals, owing to the liability of setting up electric action, which might destroy the more easily corroded metal in a very short time.

These stays must have a cross section of $\frac{390}{25000} = .0176$ sq. in., but as the ends will have to be screwed, which will reduce the effective area of the rod, $\frac{3}{16}$ in. rods should be employed, which have an area of $.0276$ sq. in.

These stays will reduce the stress coming on the end seam

* *Note.*—The end plate being $\frac{5}{64}$ in. thick does not make any material difference to the strength of the seam, as the weakest plate must always be considered.

GENERAL PRINCIPLES OF BOILER DESIGN, ETC.

by $390 \times 7 = 2730$ lbs. Having now explained the reason for and the method of calculation employed to find the bursting pressure of boilers, it now remains to state the various calculations in the form of algebraic formula.

π = the ratio of the circumference of a circle to its diameter = 3·14.

r = the radius of any circle. D = the diameter.

T = the thickness of shell in inches or fractions of an inch.

BP = bursting pressure. $WP = \dfrac{BP}{F}$.

WP = working pressure.

F = factor of safety, usually 8, 10, or 12 for boilers.

S = strength of material: 50,000 lbs. for steel, 45,000 lbs. for iron, and 25,000 lbs. per sq. in. for copper.

K = coefficient, dependent upon the type of riveted joint employed, being 100 where solid drawn tubes are employed for shells, or the figure given in the last column in the tables of riveting given on p. 8.

Then the area of a circle = πr^2.

The circumference of a circle = πD or $2\pi r$.

The thickness of shell plates for a boiler in inches—

$$T = \frac{D}{2} \times WP \times F \times \frac{100}{K} \times \frac{1}{S}.$$

The formula for finding the bursting pressure of a boiler is

$$BP = T \times S \times \frac{K}{100} \times \frac{2}{D}.$$

To find the working pressure divide the above result by F.

The above formulæ hold good for externally pressed cylinders, such as flues for Cornish boilers, *but care must be taken that the said flues are perfectly circular*, and they should not be longer than once and a half the diameter, unless supported by stiffening rings or by cross tubes, which must be thoroughly well brazed or silver-soldered in.

It is always advisable, however, to make flues with a higher factor of safety than the rest of the boiler. Thus if 12 is used for F in the case of the shell, 15 should be used for the flue.

Steel or iron boilers have to be provided with a little extra thickness beyond that found by the above calculations to allow for corrosion, as otherwise the working pressure would have to be reduced very shortly after the boiler was put to work. It is difficult to fix an exact amount to allow, as the corrosion depends upon many factors, amongst others being the nature of the water used in it, the manner of heating, the amount of use to which the boiler is put, and the care taken with it, more especially when not in actual use.

On laying up a boiler it should either be filled right up with water that has been well boiled, or carefully emptied and thoroughly dried by the application of heat, care being taken not to damage the plates by overheating. In either case, all apertures should be carefully sealed to prevent the ingress of air.

Usually about $\frac{1}{16}$ of an inch is allowed for deterioration, and, of course, it makes no difference whether the boiler is large or small, thick or thin, the life of the boiler at full working pressure will be gone when this extra thickness of metal has been corroded away, even if, as often happens, this corrosion takes place in isolated spots or ridges (technically known as " pitting " or " grooving ").

Large boilers are often used for a considerable period at gradually reduced pressures till they are absolutely unfit for further service and finally come down to being used as tanks.

Model boilers that a man cannot get inside to carry out a thorough inspection should periodically be opened up by taking one end out, and thoroughly inspected, and if corrosion shows to any extent the part affected should be replaced with a new plate, or if the defect be general it is advisable to scrap the whole boiler.

The writer of this part of the book has been called in to inspect several boilers after explosion, and has a perfect horror of anything likely to lead to such an event, as even a boiler containing 1 pint of water can contain sufficient energy to kill several persons if they happen to be standing round it.

Soft solder should never be depended upon to hold a joint together, its only admissible use being to render joints watertight after they have been securely riveted together. There is no harm in this, as it will act as a fusible plug in the event of low water, and prevent all fear of a boiler failing from this cause, as the solder would blow out and allow the steam to escape. Care must be taken, however, that the joints are soundly riveted, and that the solder only serves the purpose limited as above.

Folded joints, such as are used in making tin canisters, should also be avoided, as the very act of making the fold is liable to set up cracks which are next to impossible to detect.

All apertures in boilers exceeding 8 times the thickness of the plate in diameter should be strengthened by carefully riveting a stiffening ring round the edge of the hole. The ring should be at least $1\frac{1}{2}$ times the thickness of the plate, and the same width as the lap required for the type of riveting employed in securing it (see table of riveting).

Steam domes, etc., are usually supplied with sufficiently thick flanges to require no further stiffening, but man-holes or hand-holes must be protected in this manner. If there be any doubt as to the thickness required, make it strong enough. The calculations of such stiffening rings are rather complex, but a man with a good engineering eye will be able to tell at a glance if it is correctly proportioned. Do not say to yourself, however, " Never mind, we have the factor of safety to depend upon." The factor of safety is intended as an assurance against hidden defects, and must not be allowed to cover defects in design or miscalculations. A hidden defect may just happen to occur in the exact spot where the error is made, and the two together may lead to serious results.

After having secured safety to the boiler, the next most important consideration is to obtain " free circulation." This, however, is rather a complex subject and cannot be treated in this small handbook.

It has, however, been thoroughly thrashed out in an article on " Small Power Boilers " which appeared in *The Model Engineer* of 17th and 24th April and 1st May 1913.

The boilers illustrated in this book are all that can be desired in this direction. Care must be taken always to avoid the use of very small tubes either surrounded by or filled with water, fewer tubes of larger diameter being more efficient, although presenting less heating surface, as the circulation of the water, or the products of combustion, as the case may be, are less constricted, and consequently the hot gases have a better opportunity of transferring their heat to the water.

Water tubes less than $\frac{3}{8}$ of an inch should rarely be used, and these should not be longer than 7 in. ; $\frac{1}{2}$-in. tubes 12 in. ; $\frac{5}{8}$-in. may be 20 in. long, if they rise at an angle of 30 degrees —if vertical they may be 50 per cent. longer.

Flue tubes are better avoided in small model boilers, as they have to be too small to admit of combustion continuing within them. If they have to be used, nothing under ⅝ of an inch should be employed, and these not exceeding 8 in. in length. Three-quarters of an inch should really be regarded as the practical limit of an internally heated tube.

The steam and water capacity of a boiler will depend upon the type and the internal arrangements for heating surface. A Cornish boiler is a large one for its power, and has a large water and steam space in proportion to its heating surface; but unless fitted in brickwork or otherwise cased, having large radiation surfaces, it is very inefficient.

They certainly have more reserve power than other types of boilers, owing to the large quantity of water they contain. For this reason they take longer to get up steam. The only place where they are really advisable is for workshop or dynamo driving (especially if set as shown in fig. 3B), but the flue must be fitted with a number of water tubes and a very powerful paraffin blow-lamp used to fire them. Vertical, marine, and locomotive boilers with a number of small flue tubes have a larger heating surface in proportion to their size, and so contain a smaller quantity of water, but get steam up more rapidly. In addition, they are more powerful weight for weight than the first-named kind, and therefore when made in copper are very much less expensive to build than Cornish and Lancashire and vertical boilers of the single central flue type. Water tube boilers have a still larger heating surface in proportion to size; they hold less water, and steam more rapidly. This, however, does not apply to the type of boiler known as the "Smithies," which has come into use for locomotives to such an extent since the first edition of this book. It has a much larger range of water than the ordinary loco.-type boiler, and therefore is an admirable kind of generator for such work.

The heating surface of a boiler consists of all surfaces exposed to the heat of the flames and hot gases. The firebox contains the most effective heating surface * in the boiler, and where possible should be surrounded by water and augmented by putting in cross water tubes. The combustion chamber and small tubes in a return-tube marine boiler (see

* This is more especially the case where methylated spirit is used as fuel.

GENERAL PRINCIPLES OF BOILER DESIGN, ETC. 17

page 43) increase the heating surface considerably and enable the length of the boiler to be curtailed. In the vertical boiler the firebox is circular with either a flat or domed top, surrounded by water, the bottom being open and provided with a grate. One central flue passes from the top of the

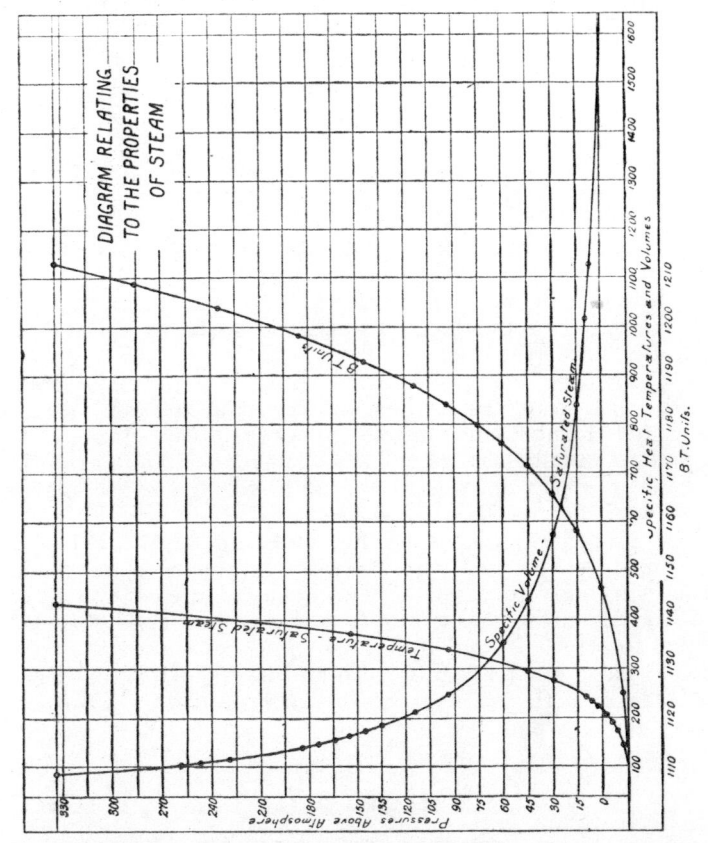

DIAGRAM RELATING TO THE PROPERTIES OF STEAM

firebox through the water and steam space, and crown plate, to chimney. Sometimes several cross water tubes are placed in the firebox inclined to the horizontal as much as possible; or in place of the central flue, a number of small flue tubes may be put in to carry the hot gases from the firebox to the

6—2

chimney, thus greatly increasing the heating surface. Boilers of this type are used on fire engines and steam launches.

The loco.-type boiler has a firebox with flat sides and slightly curved or flat top. There is a water space on all four sides and on top, the bottom being open and fitted with a grate when coal, coke, or charcoal is used. From the upper half of the front of the firebox a number of flue tubes run through the barrel of the boiler to the smokebox, these tubes being surrounded by water. Sometimes water tubes are placed in the firebox. The new model locomotive water tube generator, which is so suitable in small scale engines ($\frac{3}{8}$ to $\frac{3}{4}$ in.), for use with a methylated spirit fire (ordinary lamp), consists of two parts, the boiler proper and the outer shell. The latter is shaped exactly the same as the shell of an ordinary loco.-type boiler, except that it need not be made steam-tight and the inner barrel (the boiler), which is of smaller diameter, is slung inside the outer shell and has several water tubes fitted to its underside. The space between the two forms the flue.

Water tube boilers consist of a large number of small tubes with water inside them, connected at the bottom with a water chamber and at the top with a steam drum. The tubes are usually inclined and should never be quite horizontal. The flames play all round the tubes, which present a large and very effective heating surface.

In all boilers using solid fuel sufficient depth should be allowed for the fuel. The average evaporation of model boilers is from 1 to 3 cub. in. of water per minute per 100 sq. in. of heating surface, according to fuel used and type of boiler, equal to between 3 and 10 lbs. per sq. ft. per hour.

To enable our readers to calculate the theoretical efficiency of steam and other similar problems, a diagram of the properties of steam is given on page 17.

CHAPTER II

STATIONARY BOILERS

PERHAPS the most simple and effective boiler to drive a small engine is made from a piece of copper tube with a flat disc fixed into each end and externally fired, with a methylated spirit lamp underneath, as shown in fig. 1.

To make such a boiler, procure a piece of best drawn copper tube, say 3 in. diameter and about 10 in. long, and about $\frac{3}{64}$ in. thick. The end plates may be $\frac{1}{16}$ in. sheet brass, filed or turned circular, and made a tight fit inside the tube. Then cut two strips of copper $\frac{1}{4}$ in. $\times \frac{1}{16}$ in. about $9\frac{1}{4}$ in. long, bend them into a circle, and rivet to the inside of tube $\frac{1}{8}$ in. from each end. The end plates can next be put in, and the projecting rim of tube flanged over by lightly tapping with hammer. The ends of the tube should first be made red-hot and cooled by plunging them into water, so as to " anneal " or soften the

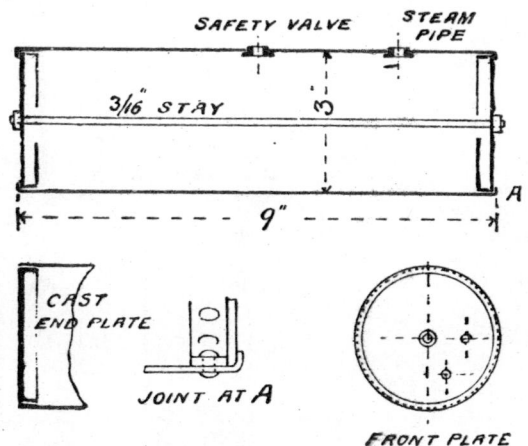

FIG. 1.—Simple horizontal boiler.

metal. Before the end plates are fixed, two or three holes should be drilled through the top of boiler at convenient places, in which can be soldered cast brass bosses with holes through centre, tapped to take the several boiler mountings or fittings.

Before fixing the rings or end plates, the surfaces in contact should be well cleaned by scraping, so that the silver solder will be sure to run all round the joint. A gas blow-pipe, or blow-lamp, will give the heat necessary. Clean off all the superfluous flux with sulphuric acid and water. In the front plate two holes can be drilled and tapped to take two small gauge cocks, one just above the centre line and the other about $\frac{3}{4}$ in. below, to show high and low water levels. The end plates should be tied together with a piece of $\frac{3}{16}$ in. copper

20   MODEL BOILER MAKING

rod screwed at each end, passing through the centre of each plate with nuts outside.

Another way of making and fixing the end plates would be to get two circular castings with a flange about $\frac{3}{8}$ in. wide, like the stamped lid of a cocoa tin. Each casting may be about $\frac{3}{32}$ or $\frac{1}{8}$ in. thick, and should be turned on its front face and rim to fit tight inside the boiler. It should then be riveted here and there or screwed through flange, the joint being made steam-tight with silver solder. No central stay will be required.

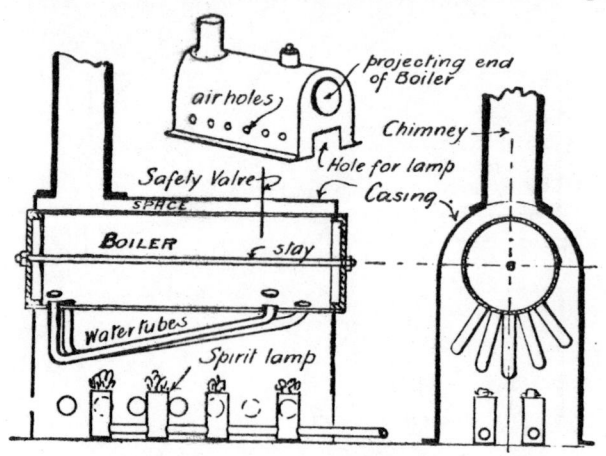

FIG. 1A.—Simple horizontal boiler with water tubes.

Stamped copper ends can now be easily obtained, and these can be riveted or silver-soldered as required.

Greater heating surface and a better circulation may be obtained by introducing some water tubes. These may be arranged in the simplest manner possible, as shown in the accompanying fig. 1A. To prevent draughts of air blowing the flame aside the boiler may be cased in as shown; it should not be closed in too much or the lamp will not burn properly. If the exhaust steam is not used to induce a draught the chimney should be large in diameter as shown.

An example of a Cornish pattern boiler 8 in. diameter and 28 in. long, with internal flue 4 in. diameter, is shown in fig. 2. To make the shell from sheet copper $\frac{1}{16}$ in. thick, a piece 28 in. long × $25\frac{7}{8}$ in. wide including $\frac{5}{8}$ in. for lap will be wanted.

STATIONARY BOILERS 21

It can be bent round a cylinder of wood about 6 in. diameter. For the diameter and pitch of rivets see table on p. 8, so $\frac{9}{64}$

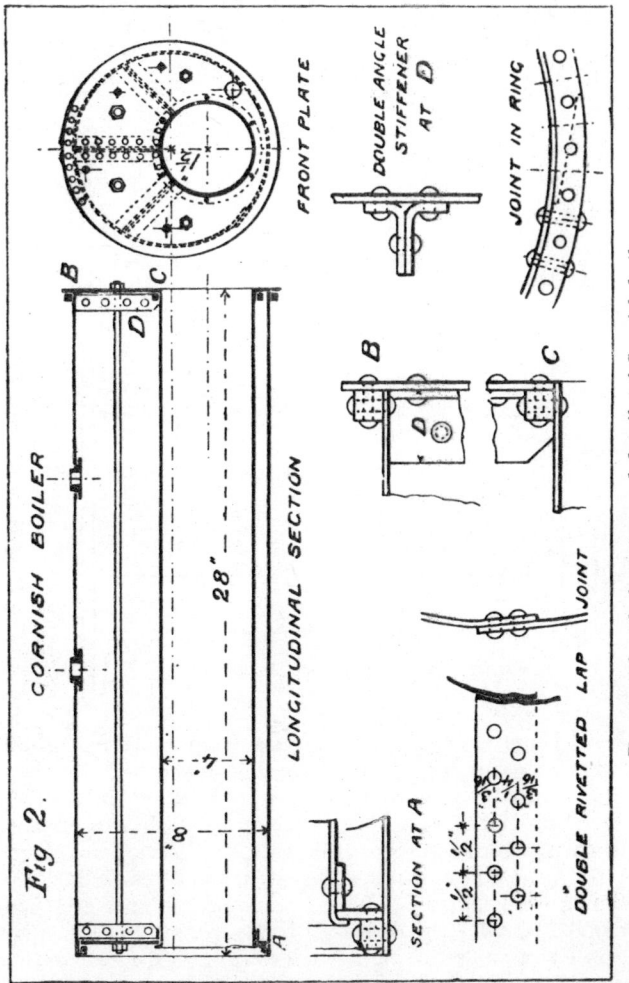

Front and end elevation and details of Cornish boiler.

in. diameter rivets will do for this boiler. It will be necessary to sweat all the joints with tinman's solder to ensure the boiler being steam-tight. A piece of copper tube 4 in. diameter may be used for the flue, but if preferred it may be made from the sheet in the same way as the shell. The front and end plates should be slightly thicker than the shell, say $\frac{3}{32}$ in., the former being $8\frac{3}{4}$ in. diameter, and projecting $\frac{3}{8}$ in. all round shell for external riveted joint presently to be described. Holes will also be required for water gauge, two gauge cocks, feed pipe, and one mud hole near the bottom, in the positions shown by the drawing of front plate.

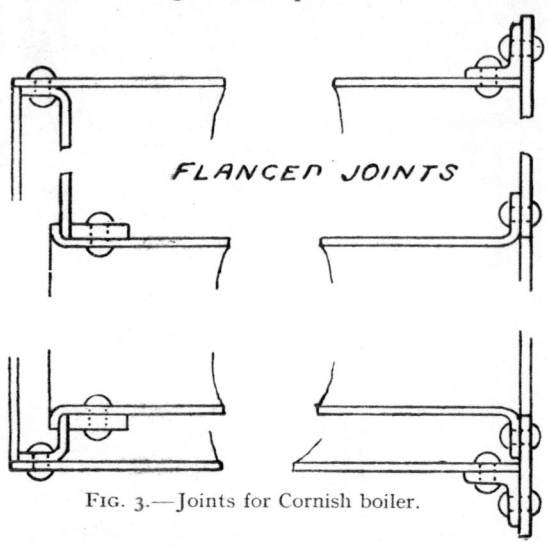

FIG. 3.—Joints for Cornish boiler.

The usual method for connecting the front plate to the shell is with an angle ring, which should be $\frac{3}{8}$ in. × $\frac{3}{8}$ in. × $\frac{3}{32}$ in. thick ; a casting could be procured say $\frac{1}{2}$ in. × $\frac{1}{2}$ in. × $\frac{1}{8}$ in. full and turned in the lathe to proper size to fit tight on to shell. Another way would be to bend a piece of $\frac{5}{16}$ in. square copper rod into a circle that will just fit over shell ; it should be about $26\frac{1}{2}$ in. long to allow for a lapped joint being made and riveted. Drill $\frac{1}{8}$ in. holes all round radially, $\frac{3}{8}$ in. apart, and between them and at right angles drill another row of holes for riveting to shell and front plate. The flue tube and front plate, the end plate and shell, are jointed in the same way.

It will be necessary to stay the end plates by putting in

STATIONARY BOILERS 23

several tie rods and tee stiffeners made from $\frac{1}{2}$ in. $\times \frac{3}{4}$ in. angle riveted together as at D, fig. 2. The tie rods should be of $\frac{1}{4}$-in. copper rod with nuts both outside and in. About six brass studs $\frac{5}{32}$ in. Whitworth should be screwed into the ring connecting furnace tube and front plate, by which the frame of firedoor is held in place. The construction of the foregoing boiler is arranged without any flanging of the plates except at end of flue, but if preferred the plates may be fixed in place

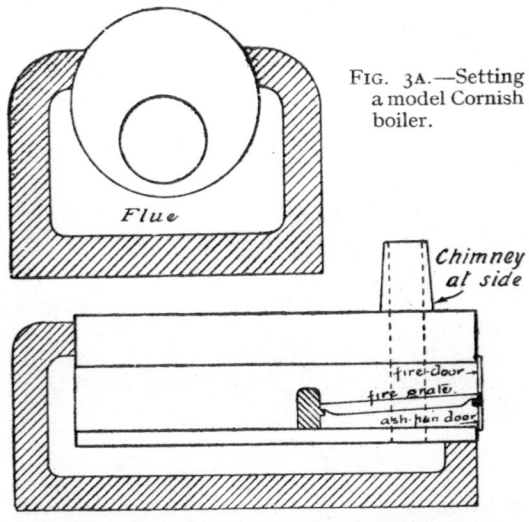

FIG. 3A.—Setting a model Cornish boiler.

Longitudinal section showing setting.

of using the copper rings as shown by fig. 3, if well silver-soldered. The safe working pressure is 25 lbs. per sq. in.

As already mentioned, this type of boiler is very large for its power, and if not set in brickwork and a good draught provided, such a generator will not evaporate more than about 1 cub. in. of water per minute (enough to run one $\frac{3}{4}$ in. $\times$ 1 in. cylinder at 300 or 400 revolutions per minute), and to increase its output should be set and fired as shown in fig. 3A. The grate, however, in this arrangement is small, and a separate furnace underneath may be used as indicated in fig. 3B. This method of firing is not so good, but if the boiler is tilted backwards slightly, so that the sediment will not accumulate over the portion directly above the fire, and the boiler is blown down frequently, no trouble should ensue. Instead of the large single " return " tube in the fig. 3B arrangement, about five 1¼ tubes might be employed with advantage.

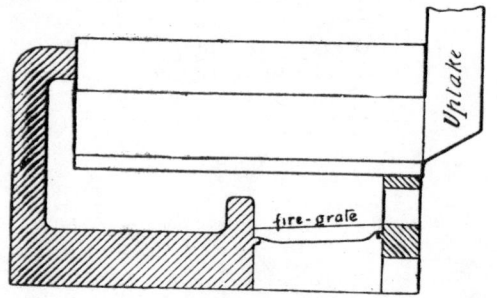

FIG. 3B.—Modification of Cornish type of model boiler.

Vertical Boilers

Vertical boilers for land use are usually made in the proportion of height equal to twice the diameter, and in place of the one central flue they may have a large number of small flue tubes or water tubes, and there are a great variety of methods of arranging the position of the tubes.

A Simple Vertical Boiler

The following is a method of making a simple vertical boiler which is intended to work at a pressure of 35 lbs. to the sq. in., and will drive an engine with one cylinder of about ⅝ in. diameter and 1¼ in. stroke, the fuel being either charcoal or half-burnt coal, and the exhaust of the engine being nozzled to $\frac{7}{64}$ in. and turned up the chimney to create a draught.

The tools necessary are as follows :—A gas blow-pipe or a petrol blow-lamp ; hand or bow drilling spindle, and a few drills from $\frac{1}{16}$ in. to ¼ in. diameter ; a hack saw ; a cross-cut chisel about ¼ in. wide ; and a strong vice, block of iron, or small anvil ; riveting appliances and a few files. The riveting appliances should consist of a block of steel or wrought iron about 1 in. by 2 in. and ½ in. thick. A snap must also be made from a piece of steel rod about $\frac{5}{16}$ in. diameter, tapered off at one end to ¼ in., the end to be hollowed out and hardened.

STATIONARY BOILERS 25

Another tool is required to hold the head of rivet while the outside head is formed; this can be a piece of bar iron 2 in. wide and ½ in. thick, 16 in. or 18 in. long. At one end of this, on the ½-in. side, drill a hole $\frac{5}{16}$ in. diameter, and let in a short

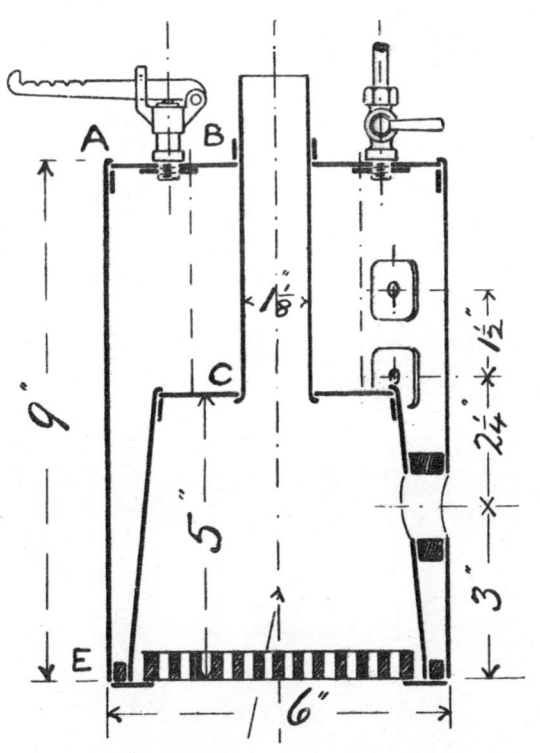

FIG. 4.—Sectional elevation of simple vertical boiler.
(*Suitable for one ⅝ in. × 1¼ in. or ¾ in. × ¾ in. cylinder.*)

piece of steel having a similar hollow in top as the snap, and projecting about ¼ in. from iron bar.

The boiler should be made from soft sheet copper. The outer shell is No. 16 S.W.G. or $\frac{1}{16}$ in. thick, and the firebox No. 14 gauge or $\frac{3}{32}$ in. thick. The conical shape of the firebox is given to make the heating surface more effective by allowing the steam to rise from the surface easily. A piece of copper tube about 1⅛ in. inside diameter will be wanted for the chimney or flue, passing from top of firebox to the outside of boiler, and projecting about 5 or 6 in. above. The top of outside and inside shell will be flat discs of copper $\frac{1}{16}$ in. thick, with a hole in centre to take the flue.

For the outside, a sheet of brass 9 in. wide and 19½ in. long

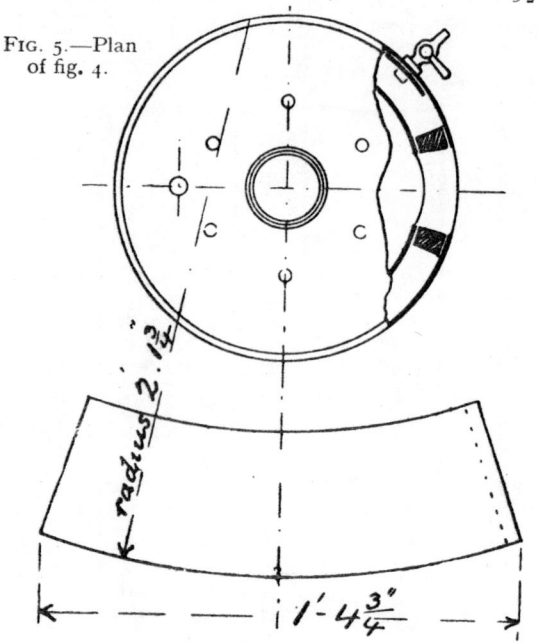

FIG. 5.—Plan of fig. 4.

FIG. 6.—Shape for forming firebox.

is required; it should be bent round a solid wood cylinder about 5 in. diameter, or a piece of hard wood about 2 in. square may have one side curved to the proper radius, and the sheet of metal bent gradually on this; a wood mallet should be used to reduce any sharp bends or corners that may appear. The two edges that overlap $\frac{9}{16}$ in. must be bent, one outwards, the other inwards, to the angle shown in drawing, fig. 7. The inside of this joint for about ½ in. top and bottom should be filed away to a feather edge to clear the top and bottom foundation ring. The joint must next be cleaned by scraping.

On the outside, mark off two lines in the centre of the lapped portion of the plate $\frac{7}{32}$ in. apart, and on each of these lines make centre dot marks at $\frac{15}{32}$ in. pitch, so arranged that the rivets in one line will fall exactly between two rivets in the other; drill the second hole from each end through the two plates while they are held together with a hand-vice and put in rivets. Then drill the rest of the holes between them and rivet up, the two end holes being left till the foundation and top rings are put in.

The rivets can be made from soft copper wire $\frac{3}{32}$ in. to $\frac{1}{8}$ in. diameter. However, as rivets can be obtained almost as cheaply as wire, most readers will not bother to make them.

The piece of copper for inside shell or firebox must be cut to the shape shown by fig. 6 in drawing, and its accuracy may be tested first by cutting out in stout paper. The lapped

FIG. 7.—Method of making longitudinal lap joint.

joint should be treated in the same way as in the outer shell, except that at the bottom it will be the outside of joint that has to be filed off flush. Holes for firedoor should now be cut through the two shells, about $\frac{1}{8}$ in. smaller all round than finished size, the centre being 3 in. from the bottom. The best way to proceed is to drill holes close together all round, the metal remaining between them being afterwards cut through with the chisel. At points where the fittings are screwed into boiler, an extra thickness of plate should be soldered and riveted to the inside, or the collars purchased with the fittings may be attached.

The ends are intended to be fixed in the same manner as those for the boiler shown in fig. 1, page 19. All the joints should be well cleaned before silver-soldering. Before soldering, flange over the projecting edge of the shell by moderate hammering all round and gradually working the edge over the discs.

A foundation ring can be made of copper $\frac{1}{4}$ in. square, and bent to the circle. Cut the copper long enough to allow the ends to be halved and riveted together. The firebox shell being slightly conical, the copper ring may be filed on the inside to fit it, or the lower edge of the firebox can be drawn in parallel by hammering (E, fig. 8).

A cast brass ring will be required as a distance piece between inner and outer shell to form the firedoor. A pattern in wood should be made curved to fit the space; it may be about $\frac{5}{16}$ in. thick, and $\frac{1}{16}$ in. wider. The ring should be oval, the smaller diameter being $1\frac{1}{8}$ in., and the larger $1\frac{1}{2}$ in.

The firedoor ring and foundation ring may be fixed temporarily to the inner shell with a few screws, so that it can be tried in its place, and any adjustments that may be found

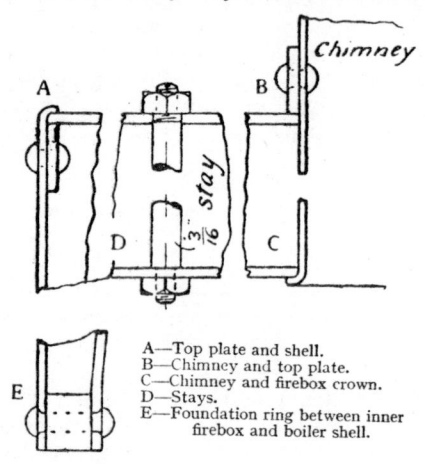

A—Top plate and shell.
B—Chimney and top plate.
C—Chimney and firebox crown.
D—Stays.
E—Foundation ring between inner firebox and boiler shell.

FIG. 8.—Details of joints for vertical boiler.

necessary can be made. The copper flue tube must be silver-soldered into the crown of the firebox before finally assembling, the lower end being slightly flanged outwards at the edge, to make a good joint with the crown of firebox. Next place the firebox in position and drill holes right through the outer shell, ring, and inner shell about $\frac{1}{2}$ in. apart; rivet up and make steam-tight with solder.

Six stays of copper rod $\frac{3}{16}$ in. diameter should be arranged round the chimney as shown in fig. 8.

Where the flue passes through top of boiler a ring of copper about $\frac{5}{16}$ in. wide and $\frac{1}{16}$ in. thick should be slipped over top of flue, riveted as shown at B, fig. 8, and the joint made good with silver solder.

The firedoor is shown in fig. 9. The hinge can be made by bending the door round a piece of wire about $\frac{1}{16}$ in. diameter, the other part of the hinge being riveted to the boiler. The catch can be made from $\frac{1}{4} \times \frac{1}{16}$ in. brass, and filed to the shape shown and screwed into brass ring.

If a solid fuel is used, an iron grate will be necessary. This can be made from iron bar $\frac{1}{2}$ in. deep and $\frac{3}{16}$ in. wide. First

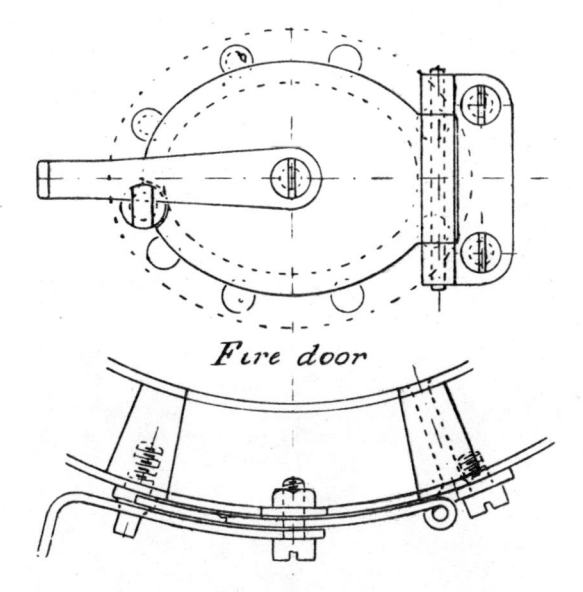

FIG. 9.—Firedoor for vertical boiler.

bend a piece to form a ring, allowing a space $\frac{1}{8}$ in. all round. The bars may be about $\frac{3}{16}$ in. apart, the ends being filed down to slip into notches filed in the ring. The grate can be supported in place by three or four brass buttons, screwed to the underside of foundation ring and projecting inside about $\frac{1}{2}$ in. It will be necessary to provide a base for the boiler to stand on, to allow air to reach the underside of grate. This may be of thin sheet iron or brass $\frac{1}{32}$ in. thick, 2 in. high, and $6\frac{1}{8}$ in. diameter inside, and should have eight holes $\frac{1}{2}$ in. diameter, equally spaced, to admit air. The boiler can rest on three or four angle pieces $\frac{1}{2}$ in. wide, riveted to the inside a little below the top.

The fittings necessary for this boiler are a safety valve $\frac{3}{16}$ in. diameter inside, steam-pipe and tap, two small taps for testing water level, and a screwed plug which can be removed when the boiler is to be filled with water. The chimney may be fitted with a turned brass cap.

When the boiler is complete it can be tested very well with a pneumatic tyre pump by substituting a suitable nipple for one of the small taps, the safety valve being plugged, and if previously filled with water any defects in the joints will show themselves after a stroke or two with pump. The points where leakage occurs can be marked and run in with soft solder by means of a good hot soldering iron.

Multitubular Vertical Boilers

Model vertical boilers of the multitubular type depend for their design upon the fuel employed. Where solid fuel such as coal, coke, or charcoal is used, a water space firebox is absolutely essential, as the side heat is considerable; but this is not so in the case of spirit, gas, or oil firing. Fig. 9A shows a small boiler having about 65 to 85 sq. in. of heating surface which can be made out of solid drawn tube 4 in. diameter with castings for the ends. The latter can be made from one pattern, which should have movable bosses and coreprints for the chimney flange and safety valve plug, on the smokebox crown and top tube plate respectively. The cast ends should be turned in the lathe, and those which are subjected to pressure should be riveted to the shell after being tinned. The whole may then be sweated steam-tight. The flame guard may, of course, be formed out of the same piece of tube as the shell, but the cheaper method will be to roll up a sheet of Russian iron or sheet steel and punch air holes in it. The boiler would run a $\frac{3}{4}$ in. × $1\frac{1}{2}$ in. horizontal engine at about 20 to 30 lbs. per sq. in. and 300 revs. per minute, or a smaller engine at higher pressure, the safe limit of which would be about 60 lbs. to 80 lbs. according to the workmanship and exact plate thickness employed. This boiler will give 25 per cent. more power if fired with $2\frac{1}{2}$ in. " Primus " oil burner. In the larger sizes (above 5 in. diameter) these boilers should always be fired with oil burners, which can be arranged in

STATIONARY BOILERS 31

clusters as shown in fig. 9C. The smokebox may also be provided with a coiled pipe as superheater as indicated.

A vertical tubular boiler 9 in. diameter inside and 20 in. high to burn coal, coke, or charcoal, and to carry a pressure of 50 lbs. per sq. in., can be made from $\frac{1}{8}$ in. sheet copper with

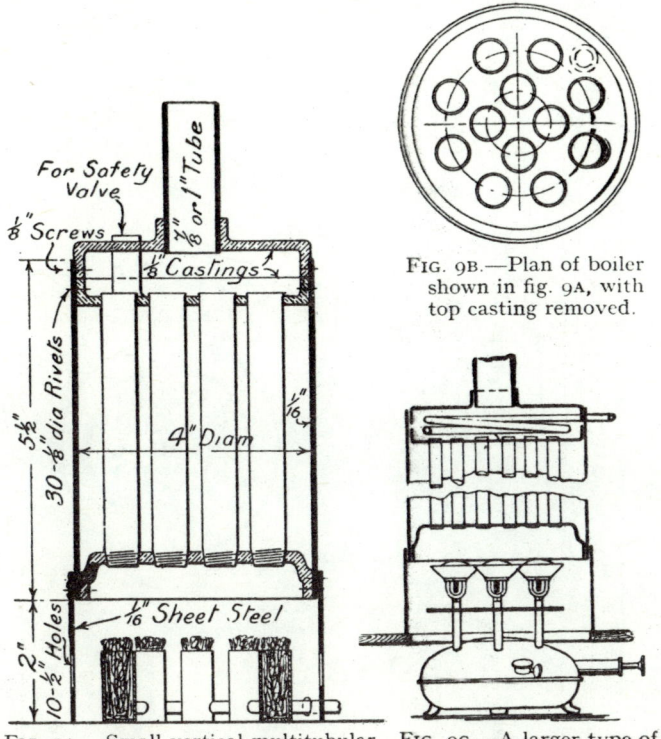

FIG. 9A.—Small vertical multitubular boiler.

FIG. 9B.—Plan of boiler shown in fig. 9A, with top casting removed.

FIG. 9C.—A larger type of multitubular boiler with Primus stove.

a double-riveted butt-jointed main seam, and as shown by fig. 10. The crown plates of the firebox and shell will be $\frac{5}{32}$ in. thick, and are flanged all round the outside to a depth of $\frac{3}{4}$ in. for riveting to the shell. There will be 18 solid drawn copper tubes, $\frac{3}{4}$ in. diameter, about $10\frac{1}{2}$ in. long. In the centre of crown plate a bolt or tie rod $\frac{3}{8}$ in. diameter may be placed

32 MODEL BOILER MAKING

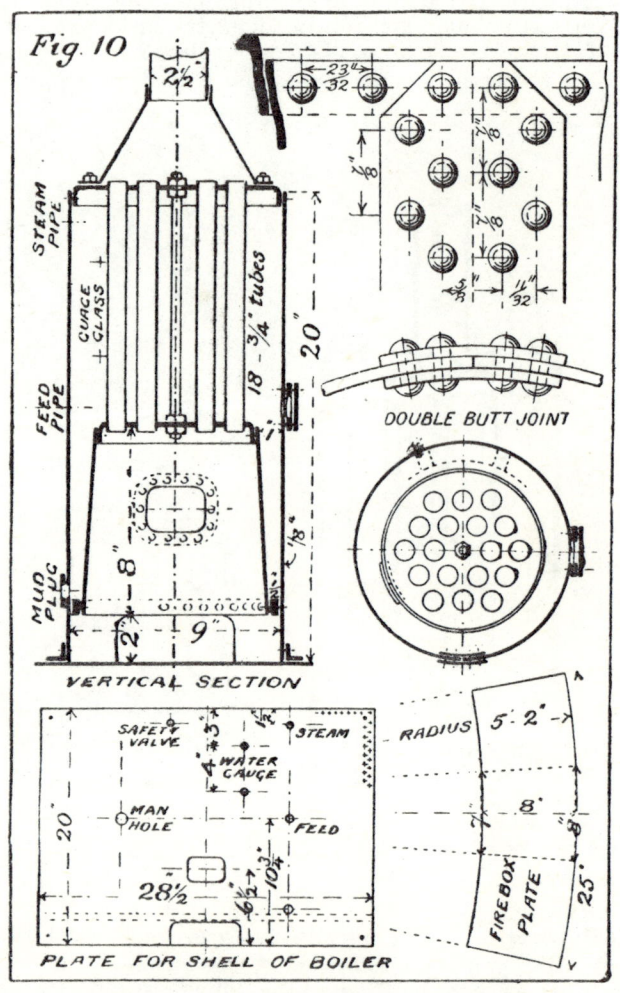

Vertical multitubular boiler for solid fuel.
(*Heating surface* 623 *sq in.*; *maximum evaporation* 7·5 *cub. in. per minute.*)

with nuts inside and out. The plate for the shell will be 20 in. × 28½ in. The two edges butting together should first be filled up straight and square. Then mark out the position of the firedoor, the hole being 2 in. × 3 in. with rounded corners, also the opening below the foundation ring to admit air to the underside of grate, and the positions of manhole, feed pipe, gauge glass, steam pipe, etc.

Previous to cutting out any of the holes, the sheet of copper must be rolled into a cylindrical shape, and when perfectly cylindrical, and the two butting edges close together, cut two strips of copper 20 in. long, 2 in. wide, and $\frac{3}{32}$ in. thick for the inside and outside cover plates. They should be curved to fit one on the inside and one on the outside of the joint.

Drill holes $\frac{3}{16}$ in. diameter at the four corners of the shell $\frac{3}{4}$ in. from each end and $\frac{11}{32}$ in. from the joint. Also drill holes to correspond in the cover strips, and bolt one on each side of joint with $\frac{3}{16}$-in. bolts. The inside strip must be shorter so as to clear the flange and foundation ring. When bolted up, the joint should be drawn up tight. Then mark out the two rows of rivet holes, each side of the joint, and drill them for $\frac{3}{16}$-in. rivets. Take the strips off, and clean off the burr and slightly countersink the holes on the outside of the strips. Then clean all the surfaces in contact, before riveting up, and " tin " them with solder to ensure a tight joint.

To strengthen the manhole a cast brass ring may be used turned true and flat on the outside. This should be riveted to the shell with six rivets, and between these, six $\frac{3}{16}$-in. studs should be screwed and riveted over on the inside of the shell. The cover for manhole should be brass $\frac{3}{16}$ in. thick and turned true and drilled to clear the studs. The joint can be made tight with red lead.

The centres of the rivet holes, $\frac{9}{32}$ in. diameter, round the top of shell and firebox must be $\frac{5}{16}$ in. from edge of plate and $\frac{23}{32}$ in. apart.

The firebox shell being conical, the correct shape must be set out full size on a plate 10 in. wide and 25 in. long, as shown further on in the chapter on setting out plates. It should be butt-jointed like the shell. The tubes should fit the holes as tightly as possible, and be expanded into the top plate and firebox crown. A brass casting may be used for foundation ring, or it may be bent up as described for the boiler on page 27, and after filing up and being affixed temporarily, the firebox can now be tried in its place to see if the firedoor ring is a good

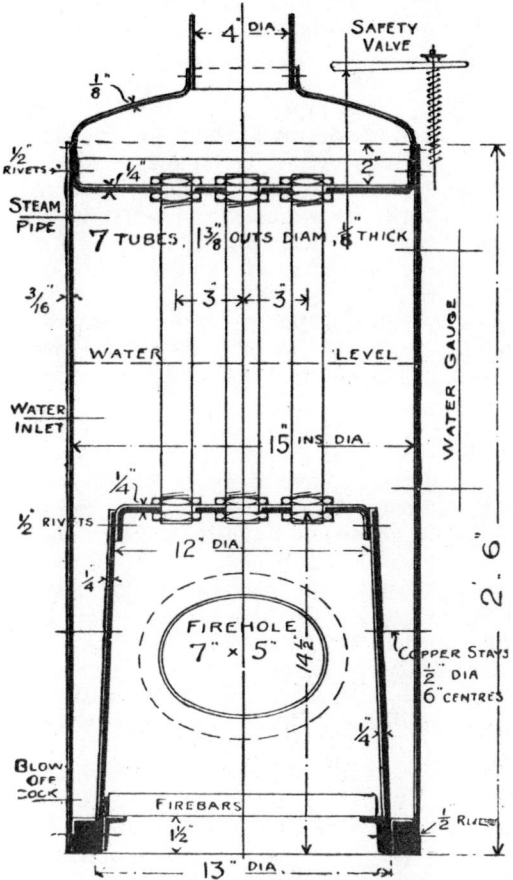

FIG. 11.—Vertical boiler for coal firing.
(*Evaporation from 9 cub. in. per minute.*)

fit against firebox when the latter is central with the shell. Next see that the holes in the crown plates for the tubes are opposite each other, and mark off the inside of firedoor on firebox and drill it out. At this stage the ⅜-in. stay bolt should be tried in position and the inside nuts adjusted to the right distance between the plates.

When the firebox is riveted in place ($\frac{9}{32}$ rivets, $\frac{23}{32}$ in. pitch) the tubes may be fitted. They should be $\frac{1}{4}$ in. longer than the distance from outside of crown plate to inside of firebox, and be annealed. The firebox ends may be slightly tapered or coned by light hammering so that they will fit tight into the holes and project $\frac{1}{8}$ in.; the other ends can then be expanded into the outside crown plate, and the projecting ends beaded over. The smokebox can be made from sheet iron, or copper, conical in section, flanged at widest end to rest on top of boiler just outside the tubes. It can be held in place by three $\frac{3}{16}$-in. brass studs. The chimney should be 10 in. high. When complete, the boiler should be tested to 150 lbs. per sq. in. with warm water.

A larger vertical boiler of slightly different construction is shown in fig. 11. It may be constructed of sheet copper. It will be seen that the crown plate is sunk 2 in. below top of shell, and the flange is outside so that the riveting is more easily done, and a slightly domed and flanged plate fits inside shell, forming a neat smokebox. The tubes are secured to the plates by being screwed at each end with a fine thread and held in place by nuts, the joint being made steam-tight with red lead. The vertical joints in the firebox and boiler shell may be similar to the double-riveted lap joint before described. This boiler may be worked at a pressure of 60 lbs. As the cost of the copper alone would be considerable (without tubes and foundation ring about £8, 0s. 0d.), the boiler can be made of steel $\frac{1}{4}$ in. thick with much advantage; the tubes then may be expanded in, a larger number used, and generally the construction simplified. Double-riveted butt-joints should be used for the longitudinal seams and the rivets driven hot The seams will require caulking.

For details of riveting, butt-strips, etc., see Chapter I.

CHAPTER III

A SMALL POWER BOILER *

THIS boiler was specially designed to drive an engine capable of developing about 5-6 H.P. on comparatively long runs with slight attention. It would be suitable for driving an engine of 2 to 4 H.P., with practically no attention beyond periodically firing—or, if a suitable gas or oil lamp were fitted, it could be left to itself for hours at a time.

Of course the lower the output demanded from the boiler the less attention it will require.

The maximum evaporation should not exceed 120 lbs. per hour, nor be less than about 50 lbs.

The pressure should not exceed 90 lbs. per square inch. It will thus be seen that by varying the pressure and regulating the fire a large variation of the power developed can be made. If greater variations of output are required the heating surface and grate area must be modified to suit.

The steam drum is made of a length of solid drawn tube with flanges and steam drum, welded on either by the oxy-acetylene or electric welding process, which of course will be beyond the average amateur, but there are a number of firms who specially cater for this class of work.

The down-comers can be built of sheet steel and welded in the same manner, or could be cast in steel or malleable iron.

If it is decided to have the various portions of the boiler welded together, the plates should be cut to shape and pinned or screwed, to hold them together whilst the various parts are being welded.

All the flanges should be turned or filed to an accurate fit, and the joints are best made with very thin asbestos yarn and " Elastic Smooth-on Cement," taking care to well blacklead one of the flanges in each joint, so that it can be taken apart without damage to the jointing material or face of the flange.

The plugs shown in figs. 12A and 12B can be purchased as a commercial article, known as " 1 in. gas-flanged barrel plugs." The flanges, however, will have to be turned down, as the stock article is provided with a flange about $1\frac{3}{4}$ in. over all.

---

* The description of this boiler, together with the calculations for evaporation, etc., first appeared in the *M. E.* for 17th April 1913. If this article be studied, it will enable the maker to modify the design, within limits, to suit the power he requires.

A SMALL POWER BOILER

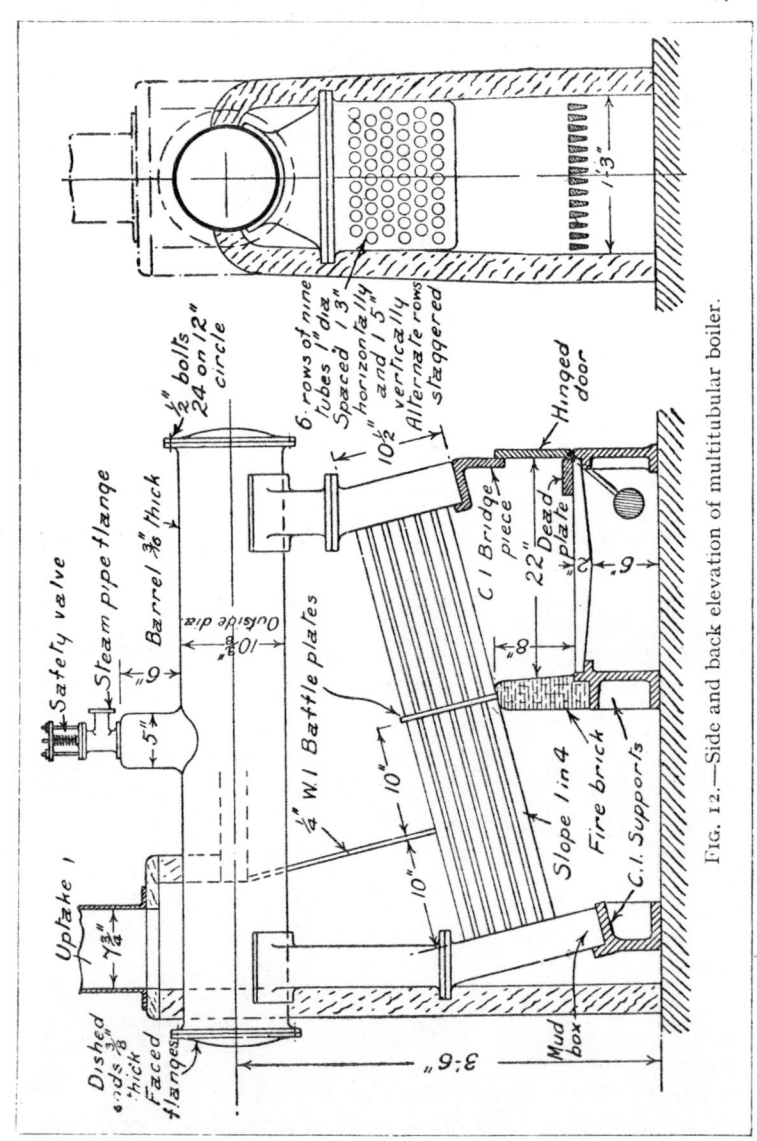

Fig. 12.—Side and back elevation of multitubular boiler.

38   MODEL BOILER MAKING

Fifteen ⅜-in. stays will have to be provided, passing right through the tubes, and six ½-in. screwed and riveted stays at the lower down-comer, as shown in figs. 12A and 12B.

The screwed plugs shown in the above figures will have to be provided first to enable the tubes to be expanded into the tube plates and afterwards to enable the tubes to be cleaned and scraped. The tubes can be expanded into the tube plates by means of the tool illustrated in fig. 12D. The body of the tool, however, will have to be lengthened so that

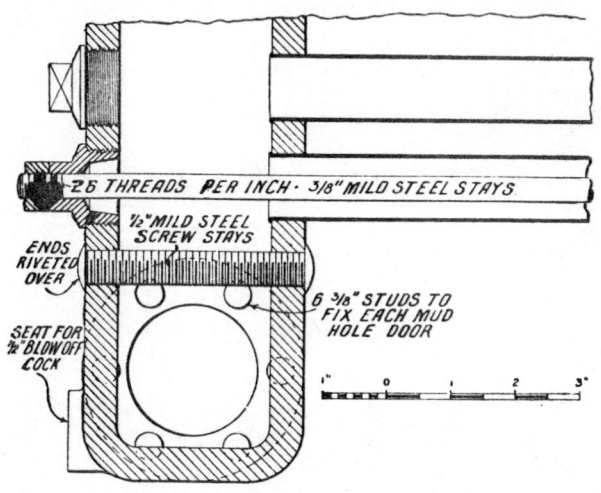

FIG. 12A.—Cross section of down-comer.

the projecting flange will rest against the outer plate of the down-comer, as the plug holes cannot be made large enough for the flange to enter.

The body of the tool consists of gunmetal or some other non-magnetic material capable of withstanding considerable strains. The form is clearly illustrated in fig. 12D, and consists of an elongated ring pierced by six holes sufficiently large to admit bicycle balls of the requisite size freely.

The bicycle balls should be about one-eighth of the internal diameter of the tube, and, of course, the gunmetal ring must be made just a loose fit in the tube.

A guide must be provided to steady the tool against the

A SMALL POWER BOILER 39

tube plate, or some other arrangement provided to keep the balls and the path round which they travel at the right position relative to the tube plate. One of the many manners

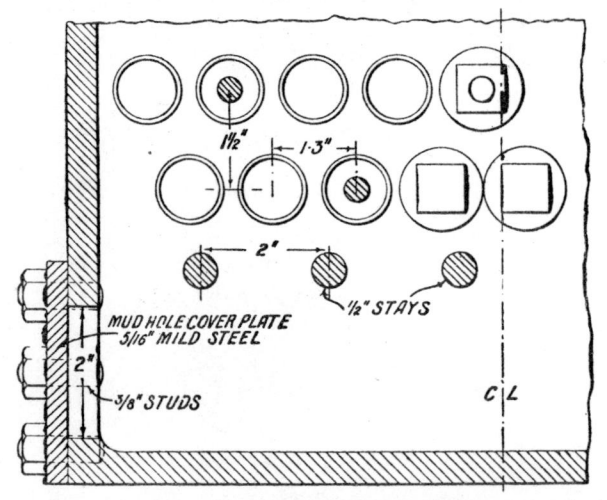

FIG. 12B.—Details of down-comer.

in which this may be done is indicated in the figure. The mandrel, which should have a taper of about $\frac{3}{16}$ in. per ft., and be about 10 ins. long, should be provided with a square at

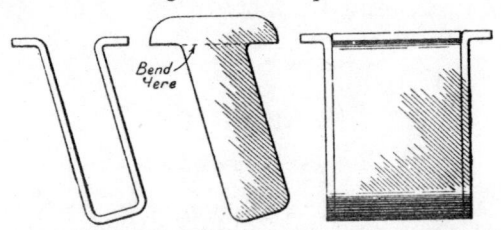

FIG. 12C.—Proposed method of constructing down-comer.

the large end made to fit an ordinary brace, and the mandrel, after being suitably tempered, should be magnetised so that when the tool is withdrawn from the tube the steel balls will be retained in position. It is advisable to pierce the thin end of the mandrel and drive in a pin short enough to enter the tube freely, but long enough to prevent the total withdrawal from the gunmetal ring, otherwise the balls would fall out every time it was removed.

The use of this appliance is exactly similar to the regulation tube expander.

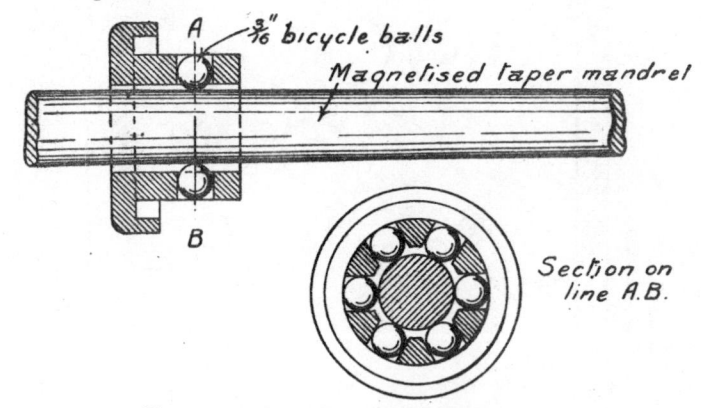

FIG. 12D.—A method of expanding a tube.

The expander should be used twice over at each end of the tube in such a manner that two rings are formed in the tubes, one near the front of the tube plate and the other close to the back. This can be easily effected by placing a washer about ¼ in. thick between the flange and the plate and subsequently using the tool without the washer.

CHAPTER IV

LAUNCH BOILERS

IN fig. 13 a boiler of the steam pinnace type is shown. The firebox extends about half the length of boiler, twenty-eight ⅝-in. brass tubes continuing to smokebox, increasing the heating surface considerably. The firebox will need staying, and the working pressure may be 50 lbs.*

* This type of boiler is difficult to construct, and we do not advise its use. We have retained the illustration as an indication of one of the many types of boiler which can be used under certain circumstances.

LAUNCH BOILERS 41

The tube plate is let into boiler shell sufficiently to form a smokebox (which may contain a superheater), a plate being fitted to the end, with a door for access to tubes.

Fig. 14 gives details of a return-tube, or Scotch boiler, much used on steamers and launches where it is desirable to economise space lengthways. For model work where benzoline or paraffin blow-lamps are used a length of twice the diameter may often be adopted with advantage. A good draught is, of course, essential, and the best means of inducing this is to use the exhaust steam from the engine. A short but large-diameter model return-tube boiler is only advisable in a very large boat, as where solid fuel is to be used a larger

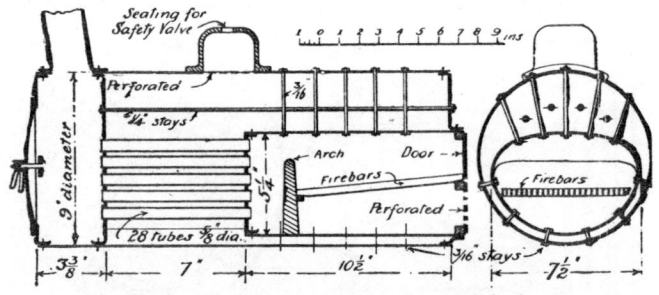

FIG. 13.—A pinnace boiler for solid fuel.

(*Heating surface about 500 sq. in. Evaporation, with induced draught, 6 cub. in. per minute. Suitable cylinder dimensions, 1¼ by 1¼.*)

grate area is obtainable. Figs. 13 and 14 show how the grate ought to be arranged for this type of boiler and for the pinnace boiler described above. A "brick" arch is necessary to prevent the air going anywhere but through the fire, and in the case of a pinnace boiler this arch, which may be a piece of cast iron, should be removable so that the back of the furnace can be cleared of ashes.

A design for a model marine boiler suitable for a boat is shown in fig. 14A. This boiler exemplifies the use of castings instead of flanged plates. The castings should come from the foundry about ⅛ in. thick, and may be slightly hammered all over to close the pores of the metal before they are turned up to fit the barrel. The boiler may be worked at 50 lbs. per sq. in., and a powerful benzoline * burner used for firing.

* Petrol may be used instead of benzoline without altering the lamp.

The furnace should contain eight or sixteen $\frac{3}{8}$-in. water tubes, arranged singly or in pairs, pitched $\frac{1}{2}$ in. apart, diagonally from right to left, diagonally left to right, and vertically, placed near to the combustion chamber end to break the flame of the lamp, as well as to provide extra heating surface. The " dry " back of the combustion chamber should be lined inside with asbestos, and the steam pipe may be coiled in front of this to form a superheater. The combustion chamber should be lined with sheet asbestos, and the end made so as to open for cleaning.

Although weight for weight, and comparing the fuel consumption,* the type of boiler shown in fig. 14B is not so efficient in large boilers ; with a powerful burner very good results

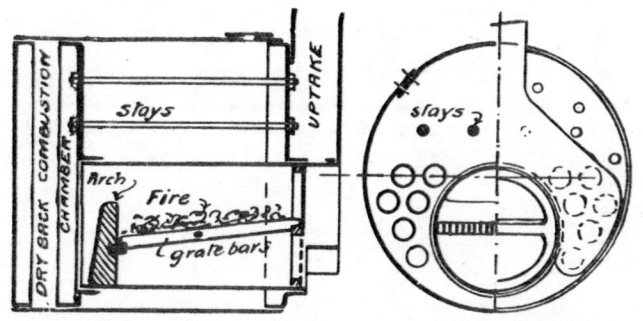

FIG. 14.—A model return-tube marine boiler of ordinary proportions.

have been obtained by its use in model racing steamers. The shell is of simple design. The boiler can be made of 5-in. and $2\frac{1}{2}$-in. solid drawn tubes, with flanged ends of $\frac{5}{64}$-in. copper plate. The water tubes should be *brazed or silver-soldered* into the flue tube. Soft solder will not stand.

Water tube boilers in model scale have a limited scope for application, and should only be used where extra high pressures are required.

The reason for this is that although they have a relatively large evaporative power, the water surface is small, so that if a large volume of steam is allowed to rise, the ebullition is very great and a large quantity of water is carried over with the steam to the engine. Fig. 15 shows a modified type

* Fuel consumption is, however, not an all-important consideration in model work.

LAUNCH BOILERS

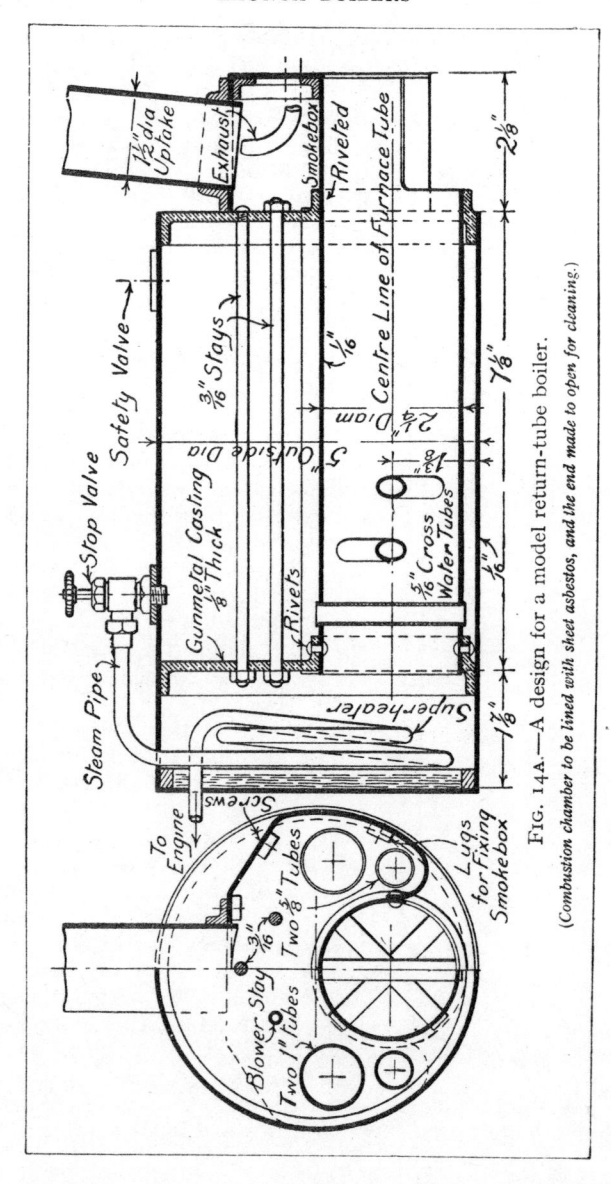

FIG. 14A.—A design for a model return-tube boiler.

(Combustion chamber to be lined with sheet asbestos, and the end made to open for cleaning.)

of water tube boiler which is designed for a pressure of 150 lbs. per sq. in., at which pressure it should supply about 1200 cub. in. of steam per minute. It would supply 2400 cub. in. of steam at 50 lbs. if it were not for the question of "priming," but priming cannot be got over, so whatever pressure is employed the boiler must not be called upon to supply more steam than that stated above. The 3-in. drum should be made of solid drawn copper tube 13 S.W.G. or ·09 in. thick. The 2 and 1½-in. tubes should be 16 S.W.G., whilst the ½-in. tubes may be 20 S.W.G. thick. Mud holes should be provided in all the longitudinal tubes as shown in fig. 15A, and all the joints should be thoroughly well brazed or silver-soldered, or rather soldered with standard silver, ordinary silver solder being too fusible for employment where the direct flame of a powerful blowlamp, such as should be employed with this type of boiler, plays directly on to the joints.

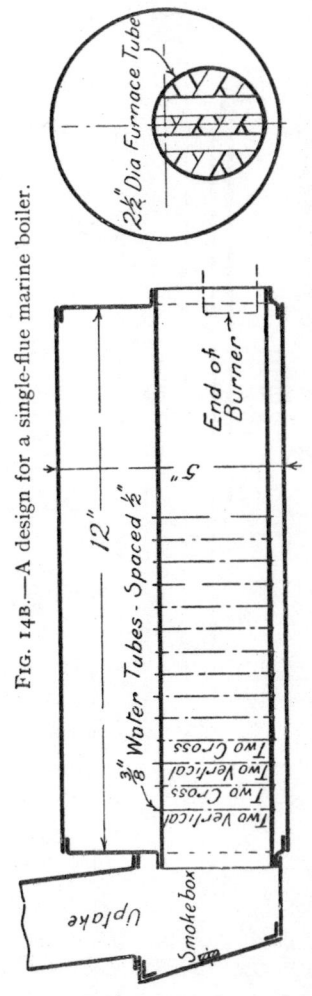

The effective heating surface of this boiler is about 280 in., and, of course, the dimensions can be varied as required, but it is not advisable to use water tubes less than ⅜ in. in diameter, and care must be taken that the aggregate area of the small tubes is less than the area of the large down-comers. These boilers work better when the end away from the burners is slightly raised. The smoke stacks (one on each side of the steam drum) should be placed near the firedoor end, as the flame will then circulate under the arch formed by the water tubes,

Fig. 14B.—A design for a single-flue marine boiler.

LAUNCH BOILERS 45

and, escaping between them will return by the passage under the upper water and steam drums.

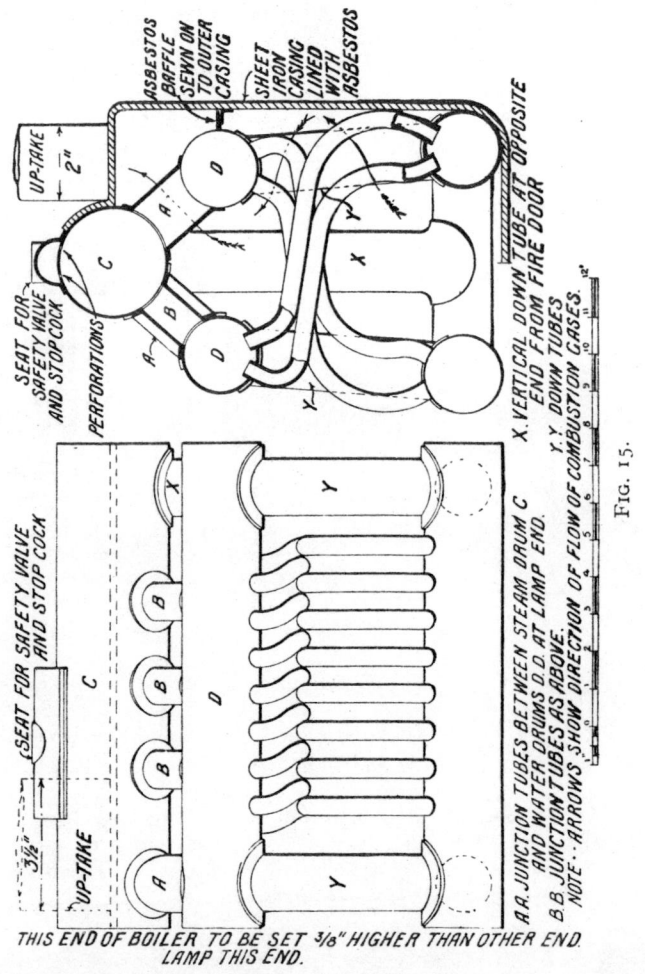

FIG. 15.

A A. JUNCTION TUBES BETWEEN STEAM DRUM C AND WATER DRUMS D.D. AT LAMP END.
B B. JUNCTION TUBES AS ABOVE.
NOTE: ARROWS SHOW DIRECTION OF FLOW OF COMBUSTION GASES.
X. VERTICAL DOWN TUBE AT OPPOSITE END FROM FIRE DOOR.
Y.Y. DOWN TUBES.

THIS END OF BOILER TO BE SET 3/8" HIGHER THAN OTHER END. LAMP THIS END.

A modification of this type of boiler is shown in fig. 15B with a 4-in. steam and water drum; this is a somewhat simpler arrangement, but is not capable of quite such a

46 MODEL BOILER MAKING

high duty as that illustrated in fig. 15. If a 4-in. drum is used, No. 10 gauge tube must be employed for the same pressure.

The boiler will have to be enclosed in a sheet-iron casing as indicated in fig. 15, which should be lined with sheet asbestos to prevent an excessive radiation of heat.

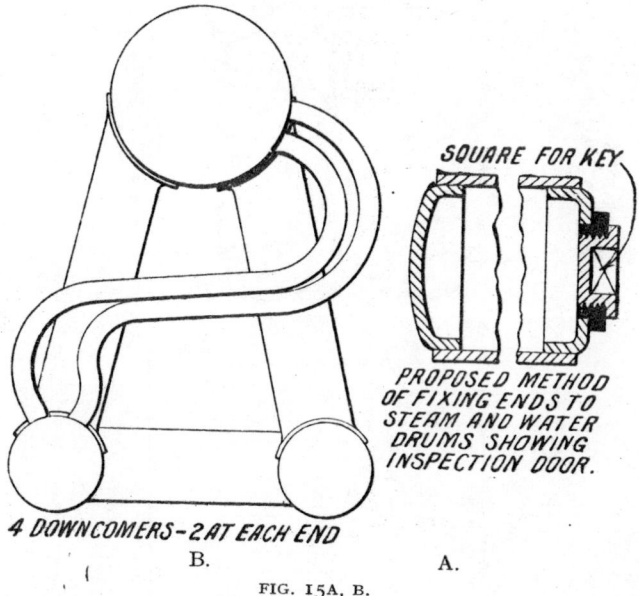

SQUARE FOR KEY

PROPOSED METHOD OF FIXING ENDS TO STEAM AND WATER DRUMS SHOWING INSPECTION DOOR.

4 DOWNCOMERS - 2 AT EACH END

B.   A.

FIG. 15A, B.

The blow-lamp for use with this boiler should be capable of burning about 1½ oz. of petrol per minute, or about 4 pints per hour.

For flash boilers see article in *The Model Engineer* for 17th April 1913 and following numbers. Other types of boilers suitable for model marine work will be found in Handbook No. 13, *Machinery for Model Steamers*.

CHAPTER V
LOCOMOTIVE BOILERS

THE boilers generally put in small model locomotives with oscillating cylinders are similar in construction to the first horizontal boiler described in Chapter II, fig. 1. With good cylinders of small size these boilers are very successful, when the flame of the lamp is well ventilated and the range of water very large. Water tubes also add to their efficiency.

Water Tube Locomotive Boilers

A considerable fillip was given to the popularity of working model locomotives soon after the first edition of this work by the introduction of water tube boilers for model locomotives. The idea consists of placing a small cylindrical boiler with water tubes on its underside inside a shell which is to all intents an external copy of that of an ordinary locomotive boiler. It would at first sight appear that the arrangement is a bad one—for various reasons, amongst which may be urged that the shell will absorb a greater part of the heat and the boiler itself get little, and that a large heating surface is impossible. These difficulties, however, are neither encountered in practice nor are so formidable as they look. The shell gets fairly warm in small engines, but not so hot as one would imagine; the painting, if done in a dark-coloured enamel, standing very well. The water tubes, which are directly over the flame of the lamp, absorb a large proportion of the heat; and the total heating surface, if it is smaller than that obtainable in a fire tube generator, is of a much more efficient character and is in the most suitable place, especially as regards liquid fuel and spirit burning engines. The barrel of the boiler is not subjected to losses by radiation as it is slung in the heated gases; and as the flues are of a large cross-sectional area (compared with a fire tube engine), the resistance to heated gases is very small. The flame is well ventilated and the draught very free under natural conditions, particularly during steam-raising if the smokebox door is left open.

Fuller details of this type of boiler are given in *The Model Engineer* from time to time. The chief points about it are: the water tubes should vary between $\frac{3}{16}$ and $\frac{5}{16}$ outside

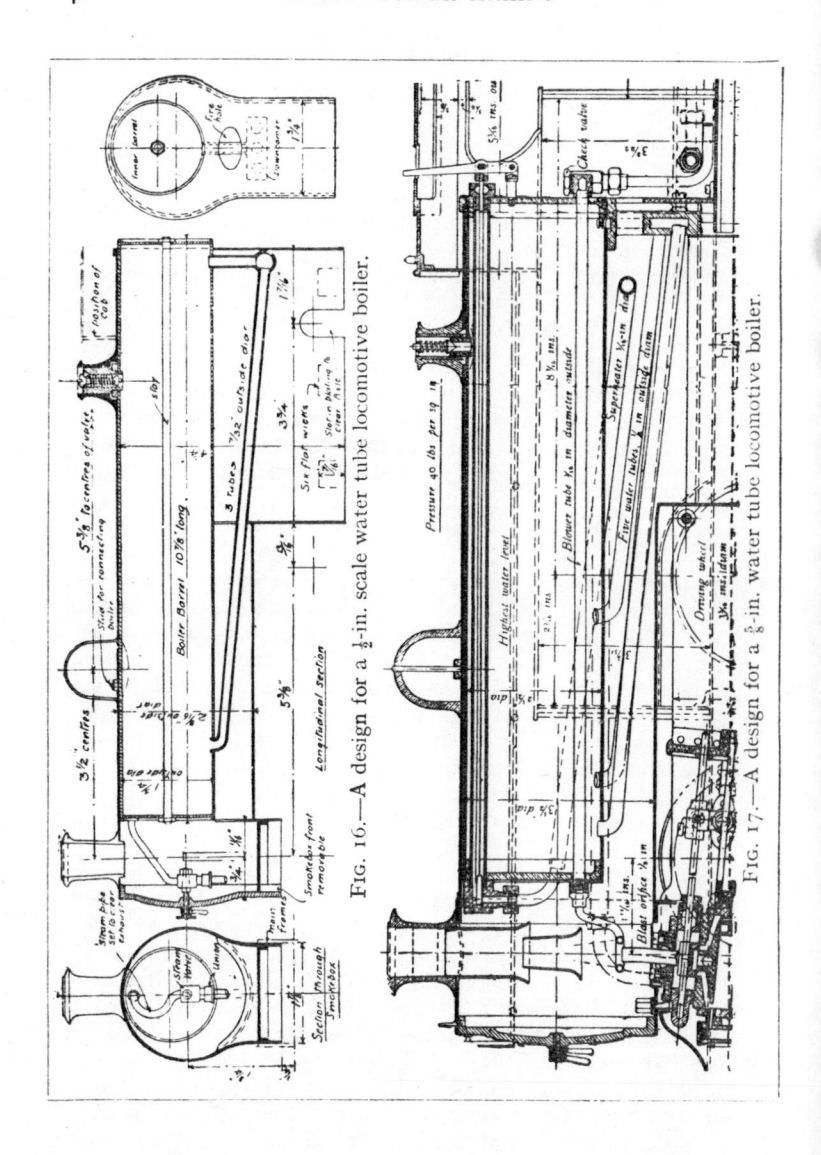

Fig. 16.—A design for a ½-in. scale water tube locomotive boiler.

Fig. 17.—A design for a ⅝-in. water tube locomotive boiler.

diameter in engines up to about $\frac{3}{4}$ in. scale, that the diameter of the inner barrel should bear a proportion of 3 to 4 to the outer shell, and that solid fuel should not be used.

As the exact proportions of the required boiler will be determined by those of the model loco. to which it is to be fitted, the subject cannot be gone into very exhaustively; therefore we give two typical designs, one for a $\frac{1}{2}$ in. scale locomotive and the other for a $\frac{5}{8}$ in. scale, the drawings (figs. 16 and 17) in each case indicating the construction. In the smaller boiler a down-comer built up of tubes is suggested, that of the $\frac{5}{8}$ in. scale steam generator being a cored casting. The inner boiler should be silver-soldered in all the main construction joints. If this is done, the ends of the larger boiler only require about six rivets in each.

Fire Tube Loco.-Boilers

For larger locomotives the ordinary type of loco.-boiler is advisable and should be used in conjunction with either oil burners or solid fuel.

Fig. 18 shows the construction of a boiler suitable for a $\frac{3}{4}$ in. scale model, in which the barrel of boiler is made from a piece of $3\frac{1}{4}$ in. drawn tube $\frac{1}{20}$ in. thick. The outside firebox plate is made from a piece of sheet copper, bent round to proper shape, and riveted to the barrel. The smokebox tube plate and front plate may be brass castings, with flanges about $\frac{1}{8}$ in. thick all over, the flange to fit outside barrel being turned, and a light cut taken over the whole plate on both sides to reduce the thickness slightly. The inside firebox is of sheet copper, the tube plate being $\frac{1}{16}$ in. thick and the rest $\frac{1}{20}$. On referring to the section of the firebox it will be noticed that the tube and back plates are widened out at top to take in the tubes; and although this entails a little extra work in the flanging, it is quite worth the trouble, as the tube space in barrel is better utilised. The back plate may also be a brass casting, with the firedoor ring cast in place.

The foundation ring is made from $\frac{3}{16}$ in. square brass, bent to shape, or a pattern can be made for a casting in brass. The sides and top of the firebox are stayed to the shell with $\frac{5}{32}$ in. brass or hard drawn copper. The stays are screwed from end to end and screwed into the inside plate from outside, and should project sufficiently at each end for a lock nut. The rivets should be $\frac{3}{32}$ in. diameter and about $\frac{3}{8}$ in. apart. It is a great advantage to have the flanged

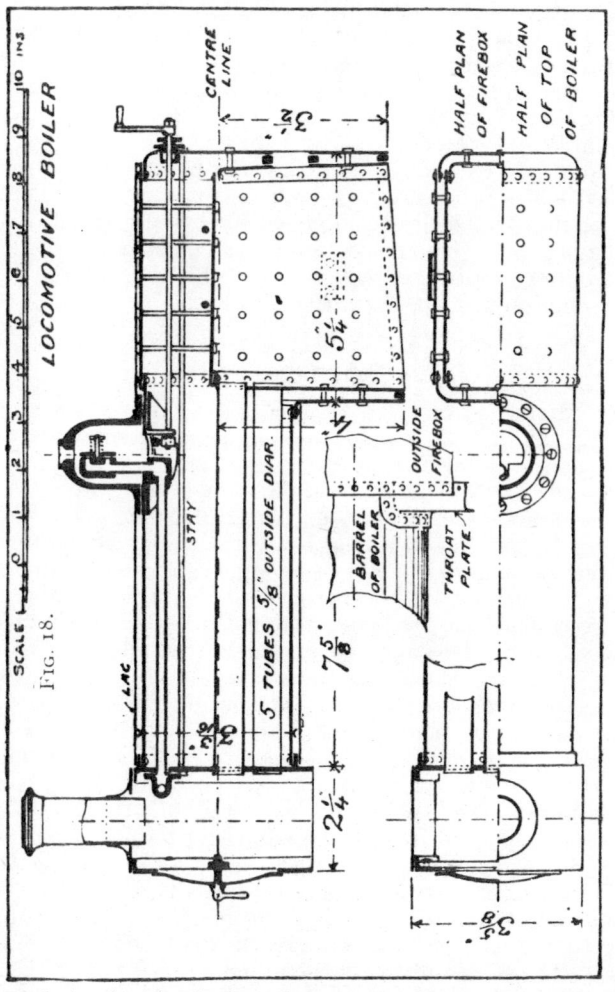

Fig. 18. — Locomotive Boiler.

joints of the inside firebox brazed or silver-soldered, a few rivets being put in to hold it together; so that when sweating the tubes into tube plate, the brazed joints will not melt.

LOCOMOTIVE BOILERS

There are five tubes of best drawn brass $\frac{5}{8}$ in. outside diameter, about 8 in. long. They should project about $\frac{1}{16}$ in.

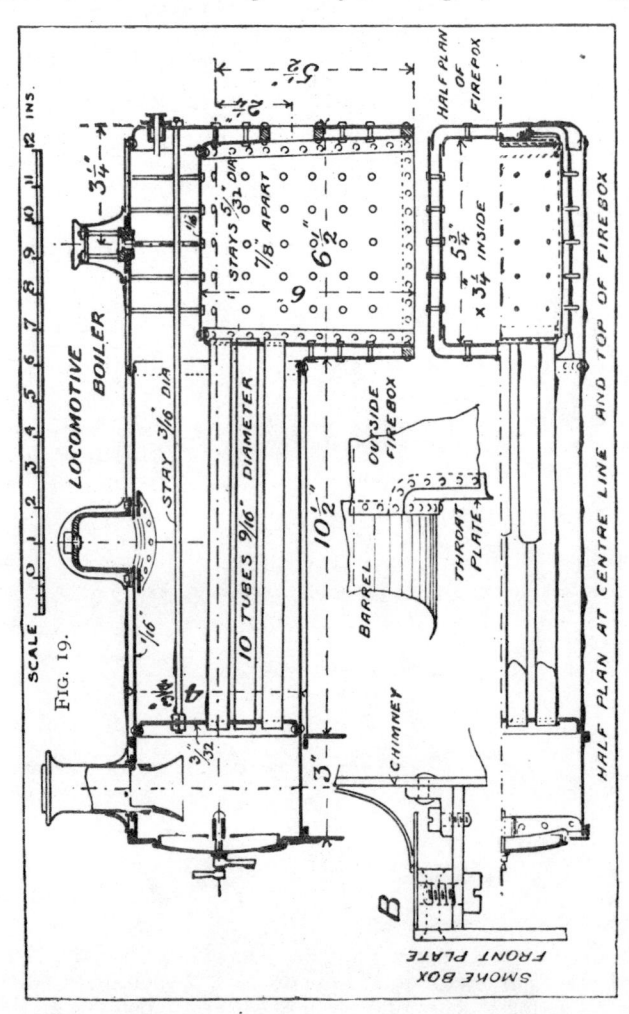

FIG. 19.

beyond firebox and smokebox tube plate, and should be flanged over. Fired by coal or a single $2\frac{1}{2}$" Primus" burner,

52 MODEL BOILER MAKING

this boiler would supply steam for a pair of cylinders $\frac{3}{4}$ in. bore and $1\frac{1}{2}$ in. stroke, with a single driving wheel 5 to $5\frac{1}{2}$ in. diameter.

Fig. 19 shows a boiler for a four-coupled express engine

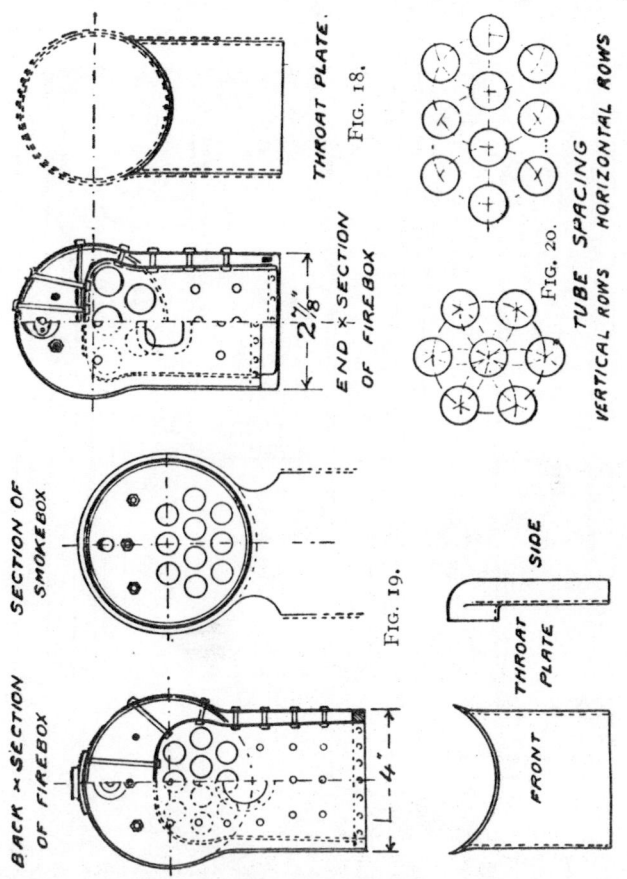

to 1 in. scale of modern type. The barrel of this boiler is $4\frac{3}{4}$ in. diameter, and $13\frac{1}{2}$ in. long, including smokebox. It is made from one piece of sheet copper $\frac{1}{16}$ in. thick and $15\frac{5}{8}$ in. wide, including $\frac{5}{8}$ in. for a double-riveted lap joint, or, if preferred, a piece of drawn tube of same diameter

LOCOMOTIVE BOILERS 53

may be used. The front tube plate is circular and flanged all round, and is riveted to the inside of barrel. The front plate of the smokebox is fixed to the shell by riveting or screwing to a piece of $\frac{5}{16}$ in. × $\frac{3}{16}$ in. brass, bent to fit outside barrel and screwed to the latter from the inside as shown at B, a similar strip being fixed at the junction of smokebox to barrel, the lagging plates on boiler and smokebox being flush. The outside firebox plate is made from a piece of sheet copper $\frac{1}{16}$ in. thick, and bent to fit outside of barrel accurately, overlapping $\frac{3}{8}$ in., and riveted. The throat plate is flanged forward to fit under side of barrel, and backward to fit between the sides of outer firebox. The back plate is flanged all round to fit inside the latter.

The inside firebox is $3\frac{1}{4}$ in. wide and $5\frac{3}{4}$ in. long inside at bottom. There are 10 brass tubes $\frac{9}{16}$ in. diameter and 11 in. long. The top of firebox is supported by direct stays in the usual way.

The sides of inside and outside firebox are stayed together with copper stays $\frac{5}{32}$ in. diameter and $\frac{7}{8}$ in. apart. Three longitudinal tie rods, $\frac{3}{16}$ in. diameter, running the full length of the boiler, stay the upper portions of front tube plate and back plate of outside firebox. All the rivets should be $\frac{1}{8}$ in. diameter and $\frac{3}{8}$ in. pitch. The working pressure is 50 to 60 lbs. per sq. in. This boiler would suit a pair of cylinders $1\frac{1}{8}$ in. bore and 2 in. stroke, with driving wheels $5\frac{1}{2}$ to $6\frac{1}{2}$ in. diameter.

CHAPTER VI

SETTING OUT PLATES, SPACING TUBES, ETC.

THE first thing of importance in making a boiler is to set out the plates correctly, as the appearance and success of the boiler mainly depend on this part of the work.

To commence with the most simple part, such as the circular shell of a boiler made from sheet metal, the width of sheet can be determined by multiplying the diameter of boiler by 3·14 or $3\frac{1}{7}$. A boiler $4\frac{3}{4}$ in. diameter would therefore require a sheet of metal 14·9 in. wide if a butt joint is used. For a lapped joint it is necessary to add the width of lap, which for single riveting is $\frac{3}{8}$ in. for $\frac{1}{16}$-in. plates, and $\frac{5}{8}$ in. lap for double riveting, the rivets being $\frac{1}{8}$ in. diameter, *i.e.* twice the thickness of plate.

For crown plates for vertical boilers and end plates for horizontal boilers which are flanged to fit inside the shell, with a rounded corner, add seven times the thickness of plate to the radius of inside diameter of shell, so that for a boiler 9 in. inside diameter and $\frac{1}{8}$-in. plates the disc of metal will be struck from a radius of $4\frac{1}{2}$ in. $+\frac{7}{8}$ in. $=5\frac{3}{8}$ in., making a disc $10\frac{3}{4}$ in. diameter. Conical fireboxes, etc., may be set out by drawing a vertical section of a cone corresponding with the vertical section of firebox or smokebox, then continue the two lines A B, until they meet at C. From that point strike a curve that will pass through upper and lower corners, and

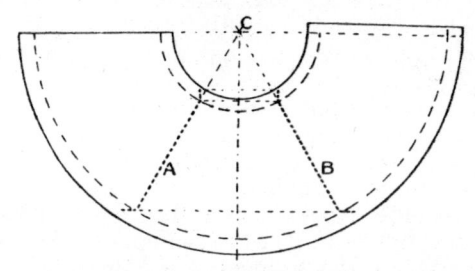

PLATE FORMING SMOKEBOX FOR BOILER FIG 10.

continue to a distance on each side equal to diameter of cone. This represents the shape to cut the metal to form the firebox, etc., adding the amount for lapped joint as before.

The plates forming the firebox of a locomotive boiler, if made correctly, require very careful setting out to get the flanges right. The tube and back plates are flanged all round except at the bottom, the sides being straight and the top slightly curved with rounded corners. Set out on the sheet of metal the exact outline of inside firebox next tube plate and back plate, then add a margin on sides and top equal to the width of flange required. In locomotive fireboxes the throat plate which forms the front of outside firebox is usually a puzzling and difficult piece of work for beginners to tackle. To get it to fit nicely at the junction of the outside firebox plate with the barrel requires great care. At this point the forward curved flange meets the backward side flange and they run together for a short distance. At this point they should

SETTING OUT PLATES, SPACING TUBES, ETC. 55

be filed off to a feather edge, filling up the wedge shape space where the side of firebox leaves the barrel. In the last two boilers described, the throat plate is about the same shape, but in the first the plate forming the outer firebox sides and top is the same width all round, and owing to the rounded corners of end plates is about ½ in. less in this dimension than length of outside firebox; and as the underside of barrel of boiler stops just in advance of throat plate it follows that the upper portion of the circumference must project backwards sufficiently to form the lap joint with outside firebox plate as shown in the adjoining sketches.

In the last boiler described the barrel is cut off square at the end, and the portion of the outside firebox plate that meets the barrel is carried forward by the amount required to form

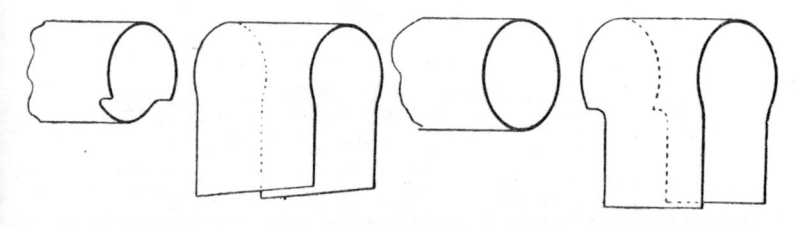

the lapped joint. It is a very good plan when a difficult piece of flanging is to be done to cut out an experimental plate in sheet lead which can easily be flanged up and will show any defects in setting out, to be corrected in the subsequent work. The outside firebox back plate is flanged forward, the upper part being circular to fit inside the circular top of outside firebox, and the sides being straight. The firebox end of the marine boiler (fig. 14) is circular with the addition of the width of flange all round, and the opening for the circular firebox must be cut smaller to the extent of width of flange all round.

Tubes in vertical or locomotive boilers are usually spaced in the way shown by fig. 20, p. 52. Set out two tubes of given diameter the proper distance apart, and draw a circle from the centre of one and cutting the centre of the other, and divide the circle into six. In this way the tube space is utilised to the best advantage, and there is the same space between each of them. They may also be arranged in vertical or horizontal rows. In small locomotive boilers, where only a few tubes can be put in, it is preferable to select such numbers as will utilise the space available most effectively. Tubes should be spaced from one-third to half their diameter apart.

The fixing of tubes in small locomotive boilers where the tube plates are $\frac{1}{16}$ in. or less in thickness, and the tubes under $\frac{1}{32}$ in., is a somewhat difficult matter. The confined space in firebox and the thinness of the plates preclude the use of a tube expander, and silver-soldering or brazing is not always possible. The exact process depends upon the method of erecting or putting together the various portions of the boiler. In either of the locomotive boilers shown by figs. 18 and 19, the front tube plate, with holes drilled for tubes, outer firebox with throat plate and back plate are first fixed in the barrel. The inside firebox is riveted up complete, with foundation ring and firedoor ring fixed with a few rivets and screws, and the joints brazed or silver-soldered, the latter requiring rather less heat. The holes for the tubes being set out, they are drilled $\frac{1}{32}$ in. less than the diameter of tube through the firebox tube plate and the same diameter as tube through smokebox tube plate, and are slightly enlarged with a taper reamer. The tubes being long enough to project beyond tube plate $\frac{1}{16}$ in. each end, they are annealed and tapered at one end to fit tight in firebox tube plate, and the other end is enlarged to fit tight in smokebox plate. Then the inside firebox should be put in position, the rivets put in, and the stays fixed. The tubes will now pass through holes in front tube plate easily, and should fit tight when the tapered ends are driven into the plates. The projecting ends can be beaded over with a small riveting hammer so that they are held firmly in place and act as stays between the tube plates. The parts having been well cleaned before putting together, the tube ends can be sweated all round with a blow-lamp or Bunsen flame.

The flanging of sheet copper and brass is done cold, and for thicknesses up to $\frac{1}{16}$ in. a beech or boxwood block properly shaped is sufficient, but for thicker metal an iron bar or block is better. The crown plate of a vertical boiler can be flanged easily on a bar of iron, say, $1\frac{1}{2}$ in. wide, 1 in. thick, as in fig. 21. The plate of copper being cut to its proper diameter, including width of flange, and well annealed, drill a hole through centre and sweat in a $\frac{3}{16}$ in. rivet with head outside. This will fit in the hole in the bar, the edge of plate projecting beyond the curved end. Then with a boxwood mallet proceed to work down the edge of plate a little at a time, turning the plate

SETTING OUT PLATES, SPACING TUBES, ETC. 57

round on its centre till the flange is just slightly turned down all round. After working round several times in this way it will be necessary to reanneal the edge of plate before the flanging is completed. The more gradually it is formed the better. If it is beaten down too much in one place it will cause buckling, as time will not be given for compressing the metal in edge of plate equally all round. The firebox plates of a locomotive can be flanged on a piece of hard wood cut to the shape of the inside of box and reduced at each end equal to the thickness of the plate. The straight portion of flanging should be worked down at the same rate as the curved corners.

Before testing a model boiler that has just been finished,

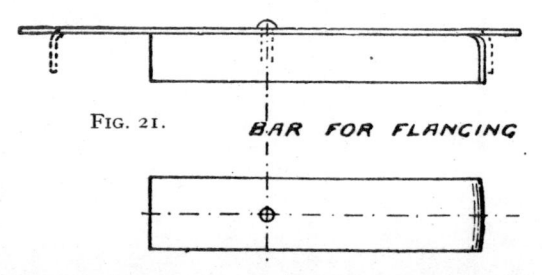

FIG. 21.   BAR FOR FLANGING

it is advisable to make sure that no small blow holes exist in the soldered joints. Stop the openings with wood or cork plugs, leaving one of the taps to blow into, put a little water in the boiler and shake it all over. Then apply a slight air pressure by blowing into the tap. If bubbles appear at one or two places, the faulty spots can be marked and resoldered without getting rid of the small quantity of water in the boiler. Try the test again, and, if all appears tight, connect up a tyre inflator to the boiler, and pump in a few pounds air pressure under water, which may discover some minute crevices. After these are all made tight, a strength test by water pressure can be made.

Fix on the pressure gauge, which should read to about three times the working pressure, and connect up the force pump with the feed pipe. Then fill the boiler up with water, giving a few extra strokes with the pump to make sure it is quite full. Then close the opening left for escape of air, and work the force pump until the pressure required is shown on gauge.

The boiler should be previously wiped dry, so that any defects can be detected, and if no bulging occurs in the flat plates through insufficient staying the test may be considered satisfactory. The hydraulic test should be twice the working pressure.

If a pressure gauge is not available, a very good substitute for testing boilers can be made, as shown by fig. 22, which consists of a piece of treblet tube 3 in. long, ¼ in. inside diameter, with a brass plug soldered into one end to screw into boiler in place of safety valve or other mounting. A piece of ¼ in. brass rod is then turned up to slide easily in the tube. At the lower end of this rod turn two or three small grooves, into which wind some lamp cotton and tallow to make it watertight in the tube; a small hole is drilled through side of tube about 1 in. from top to allow the escape of water when the pressure is exceeded. On top of brass rod solder or screw a disc of brass about $\frac{1}{16}$ in. thick and 2 or 3 in. diameter, to carry the weights. We will suppose a boiler is to be tested to 50 lbs. per sq. in.: first find the area of tube or plunger, which in this case is ·049 or nearly one-twentieth of a square inch. Then the weight to be put on top of disc will be 2½ lbs.* Instead of using a force pump, a very good method is to fill boiler with cold water and light a lamp under firebox. As the water warms up it will expand, and when the plunger and weights are raised the pressure may be taken as sufficient and the lamp put out. The steam test should be to one and a half times the working pressure. Conversely, if a pump is not available, the device shown in fig. 22 may be used as a means of testing without further apparatus. Knowing the area of the tube or plunger

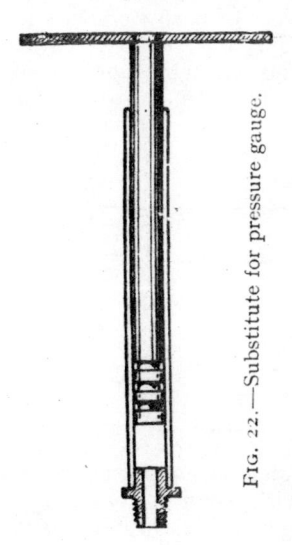

FIG. 22.—Substitute for pressure gauge.

---

* If a yard or so of fine thread be taken, and a few turns made round the plunger, so that by pulling either end of the thread alternately a rotary motion can be given to the plunger, the effect of sliding friction can be entirely obliterated.

CHAPTER VII

FUELS, LAMPS, FIRE GRATES

THERE are numerous fuels suitable for model boilers, from methylated spirit to Welsh coal, but various types and sizes of boilers will do best on one or the other kind, depending on their design, construction, and purpose.

Methylated spirit (spirits of wine) is most used, and is undoubtedly the best for boilers with heating surfaces up to 100 sq. in. A lamp suitable for the boiler described on pages

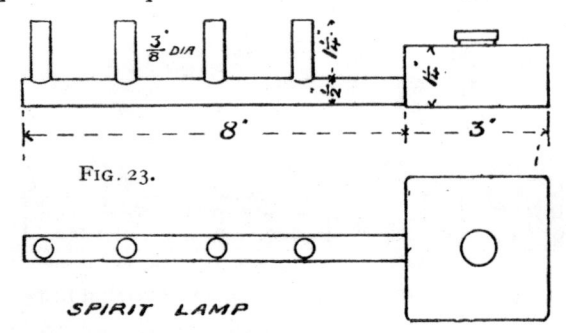

FIG. 23.

SPIRIT LAMP

18 to 20 is shown by fig. 23. It consists of a tin reservoir with a projecting tube: into this tube there are soldered three or four short pieces of tube $\frac{3}{8}$ in. diameter and $1\frac{1}{4}$ in. high; these are packed with asbestos wick. In the top of tin box a hole is made $\frac{3}{4}$ in. diameter for filling up.

Spirit burners for larger boilers are more safe and better regulated if made on the principle of the burner shown by fig. 24, in which the supply of spirits is contained in a separate reservoir away from the fire, and the quantity allowed to flow into burner under firebox can be adjusted to the amount necessary to keep up steam. Otherwise, if the burner contains the whole supply of spirits, and is close to firebox, the radiation of heat causes the spirits to vaporise and boil over, perhaps enveloping the whole model in flame.

tive boiler in fig. 19. The two end pieces have slots cut half way down. The bars are filed away to half their depth to fit tight in these slots; the distance between the bars should be about equal to their thickness. The grate can also be cast in one piece in iron from a wood pattern, in which case the strips of wood forming the bars should be tapered slightly, the widest portion being at the top.